39.03

Anatomy of a Conflict

Terre Satterfield

Anatomy of a Conflict:
Identity, Knowledge, and Emotion
in Old-Growth Forests

UBCPress · Vancouver · Toronto

09   5 4 3

Printed in Canada on acid-free paper

National Library of Canada Cataloguing in Publication Data

Satterfield, Terre.
Anatomy of a conflict

   Includes bibliographical references and index.
   ISBN 0-7748-0892-6 (bound); 0-7748-0893-4 (pbk.)

   1. Old growth forest conservation – Northwest, Pacific. 2. Sustainable forestry –
Northwest, Pacific. 3. Northwest, Pacific – Social conditions. I. Title.
SD387.O43S27 2002   333.75'17'09795   C2002-910239-1

Canadä

UBC Press gratefully acknowledges the financial support for our publishing
program of the Government of Canada through the Book Publishing Industry
Development Program (BPIDP), and of the Canada Council for the Arts, and the
British Columbia Arts Council.

This book has been published with the help of a grant from the Humanities and
Social Sciences Federation of Canada, using funds provided by the Social Sciences
and Humanities Research Council of Canada, and with the help of the K.D.
Srivastava Fund.

Printed and bound in Canada by University of Toronto Press

Set in Stone by Artegraphica Design Co. Ltd.
Copy editor: Joanne Richardson
Proofreader: Deborah Kerr
Indexer: Heather Ebbs

UBC Press
The University of British Columbia
2029 West Mall
Vancouver, BC V6T 1Z2
604-822-5959 / Fax: 604-822-6083
www.ubcpress.ca

Contents

Acknowledgments / vii

Note on Names and Methods / ix

1 Introduction: A Cultural Dialogue about Old-Growth Forests / 3

2 The Cycle of History: Public Lands, Forest Health, and Activist Histories in the American West / 15

3 Disturbances in the Field and the Defining of Social Movements / 38

4 Negotiating Agency and the Quest for Grassroots Legitimacy / 63

5 Voodoo Science and Common Sense / 81

6 Theorizing Culture: Defining the Past and Imagining the Possible / 99

7 Irrational Actors: Emotions, Ethics, and the Ecocentred Self / 135

8 A Concluding Discussion: The Triangular Shape of Cultural Production / 160

Notes / 172

References / 181

Index / 190

Acknowledgments

Among the oddities that characterize anthropological investigations is the fact that anthropologists routinely enter worlds to which they do not belong and ask a great deal of those who reside there. Happily, it is usually the researcher's discovery that people are profoundly open and generous with their time and their insights, or such was my experience. Perhaps some of this generosity was heightened in my case by the logging controversy itself as conflicts tend to drive our deepest convictions out of the closet, propelling us to lay them open for reflection and debate. Regardless, I am deeply grateful to the communities of loggers and environmentalists that have informed this work. The book is dedicated to them; I remain in awe of their ardour for forest life and politics as they make civil society the rich and complicated field that it is.

Several people have contributed to the ideas herein. Philip Bock, Louise Lamphere, and Keith Basso were influential during my graduate school years, and helped shape my anthropological sensibilities generally. Dorothy Holland's chapter readings (and superb work with her colleagues on identity and culture) deserve particular mention, as do Samuel Porter's insights into the controversy. Colleagues and graduate students at the University of British Columbia's Institute for Resources and the Environment have kept me on my toes and made teaching all the more relevant to research and writing. Support provided by Randy Schmidt, editor at UBC Press, was most welcome at several junctures. Helpful and occasionally searching readings of specific chapters were provided by Louisa Cameron, Julie Cruikshank, Melissa Finucane, Lisa Gezon, Robin Gregory, Jennifer Kramer, Charles Menzies, Madeline Mooney, Paul Slovic, and, especially, Kay Milton and Rob VanWynseberghe. Panelists at the American Anthropology Association, Society for Applied Anthropology, and the arts and humanities-inspired Community and Environment meetings offered astute comments on presented portions of the book, as did UBC Press's anonymous reviewers. Several good ideas were also provided by reviewers of Chapters 4 and 5, which appeared in a slightly different form in, respectively, a 1996 special issue on grassroots organizing in the *Journal of Social Issues* and a 1997 volume of the *Journal of Anthropological Research*. All photographs were graciously provided by photojournalist Elizabeth Feryl.

Friends and colleagues at Decision Research, especially the as yet unmentioned Sarah Lichtenstein, Donald MacGregor, Ellen Peters, C.K. Mertz, and Jim Flynn, have graciously accommodated a work in progress and have offered considerable intellectual help. Joanne Richardson offered valuable editorial suggestions; whilst Janet Kershner, Leisha Wharfield, and, especially, Mona Bronson helped prepare the manuscript with just the right blend of humour and technical care.

Invaluable financial support was provided by a Social Sciences and Humanities Research Council of Canada Doctoral Fellowship, a grant from Canada Council's Aid to Scholarly Publications Programme, and grants SBR9308246 and SBR9602155 from the US National Science Foundation. Any opinions, findings, conclusions, or recommendations expressed here are my own responsibility and do not necessarily reflect the views of the above funding agencies.

I am also indebted, for moral support during different periods of research and writing, to Gail Evans, MaryJane McReynolds, the Miller family, Madeline Mooney, Scott Slovic, Gail Wilson, the Wittman-Mabrey household, and twenty years of friendship with the Wednesday night swimming and dinner group.

Thank you, most of all, to my family near and far, especially to Mary and Peter Satterfield, for their immeasurable support, and to Sherry Burns for her wise musings on human nature in all its odd and ordinary forms.

Note on Names and Methods

It is commonplace for anthropologists to use pseudonyms to protect the identities of those who collaborated as interviewees or who were observed during the fieldwork period. The practice is more effective, however, when fieldwork is carried out in remote corners of the globe. When collaborators are proximate, the practice is less reliable: some parties may recognize (or think they recognize) themselves in the pages that follow. I have thus taken several additional steps to protect the identities of those involved. Along with the use of fictitious first names and surnames, I have changed the names of all small towns. This is because of the close association between some communities and the more prominent spokespersons who lived there. (Where applicable, I do use the names of cities, as urban areas are – by definition – more anonymous than rural areas.) I have used pseudonyms for the names of activist groups and some conferences, and, as necessary, I have limited the specificity of geographic or characteristic identifiers of smaller groups. In some, though not all, cases the speaker's title or job description is altered. I also use pseudonyms when quoting national spokespersons whom I did not interview but who appeared at local public events.

Virtually all forest-oriented grassroots activists living in Oregon and operating at the time of this study were immediate or more distant affiliates of two statewide organizations: the pseudonymous Oregon Forest Community Coalition (for loggers and timber-dependent communities generally) and the pseudonymous Ancient Forest Grassroots Alliance (for environmentalists). Other interest groups, such as those representing hunters, anglers, and ranchers, were present at the time and often belonged or offered their support to the above coalitions. But during this period of the study, the primary activist parties were organized loggers and residents of timber communities (formally known as the Forest Community Movement) and grassroots environmentalists (known as the Ancient Forest Movement).

Fieldwork interactions with groups and individuals, as well as observations gathered at significant meetings or conference events, are described more fully in the following pages. Many of the book's substantive data are provided equally by ethnographic interviews. Interviewees were drawn from representatives of the above organizations as well as from active participants dispersed across the coalitions' member groups. I spoke with both highly public activists as well as with those who worked tirelessly and quietly in tattered offices around the state. Often these conversations occupied an entire morning, afternoon, or evening, and some people were interviewed on two or more occasions. I spent a little more time interviewing people in the timber community network than I did interviewing people in the environmental community in order to compensate for the fact that the former were less familiar to me than were the latter.

Many of those interviewed are quoted at length in the subsequent chapters. If I am guilty of excessively quoting articulate informants, then that is solely because they captured so beautifully (and were representative of) what many others had to say. Nonetheless, I am careful to point out and to fully incorporate the variations on thought and behaviour among and across activists.

Anatomy of a Conflict

1
Introduction:
A Cultural Dialogue about
Old-Growth Forests

Few environmental controversies have been more dramatic than the one over the destiny of Oregon's temperate rain forests – a controversy that, in the last decade, has centred on the practice of old-growth logging and the survival of the endangered northern spotted owl (*Strix occidentalis caurina*). The battle has raged throughout British Columbia, Washington, and northern California as well as Oregon and has been aptly referred to as a "slow motion riot" (Robertson 1996, 1A). It has fuelled or coalesced two primary social movements (the Ancient Forest Movement and the Forest Community Movement), and it has resulted in such acts as the burning down of ranger stations, the spiking of trees, logging truck blockades, and countless demonstrations.

Economists, forest scientists, and historians have all worked to explain the social and ecological ramifications of reduced logging on public lands. Little has been said, however, by cultural anthropologists. Even less has been said, at least ethnographically, about the controversy's primary activist-protagonists: loggers and environmentalists. *Anatomy of a Conflict*, which is a cultural analysis of the old-growth dispute, means to remedy both these omissions. It is rooted in primary (ethnographic) and archival research conducted throughout the 1990s. Most of the ethnographic work took place between 1992 and 1996, the controversy's peak years. The years that followed have been equally rich due to the continuously emerging literature on environmental disputes.

Rather than thinking of the conflict as a contest of political, economic, and scientific forces (which it certainly is), I choose as my primary objective to draw attention to how these forces are expressed culturally in the impassioned wrangling of locally situated and politically committed loggers and environmentalists. The intent is twofold. First, I mean to demonstrate the creative means through which opposing activists, each caught in the larger web of the aforementioned forces, achieve their ends. Second, I mean to demonstrate that the forest dispute has everything to do with imagined ideal worlds, with the creative manipulation of political discourse, the assertion of moral priorities and identities, and with how activists on both sides appropriate linguistic and symbolic tools in order to promote a cultural world that reflects their quests for change.

An anecdote is a suitable place to begin. This one tells of an event that took place part way through my research period. An Oregon acquaintance invited me to speak to his middle school students about the forest controversy, which, due in part to federal policy aimed at protecting the spotted owl, had reached a feverish pitch. The stakes involved for parties on both sides were extremely high. A continuation of logging at recent levels promised ecological disaster, while a temporary suspension and subsequent decline in the annual allowable cut portended mill closures and job losses. The dispute had become, in the words of Oregon Congressman Peter DeFazio (D-OR), a "religious war." Those who "deviate from the true faith – whichever true faith – are condemned as sinners, heretics, or worse" (Porter 1999, 3).

Paring the fray down to a few succinct nuggets for middle school classroom discussion was simple enough. A more difficult task was answering the students' inquiries as to my own allegiances: Whose side, they wanted to know, was I on? It was a reasonable question: after all, I was working in the field of the spotted owl conflict, and the students' tranquil, once timber-dependent city was contentiously divided on the subject of logging. Standing mid-room I offered a neutral, though honest, reply: "I have always been drawn to these forests and think of them in protective terms, though my conversations with displaced loggers have deeply affected me. I suppose that leaves me torn between two worlds."[1] It was not the desired response, as it failed to appease the growing tension in the classroom. The usual classroom cacophony of competing voices, the clamour of books, papers, and backpacks came to an uneasy halt. It appeared they hoped I would resolve a rift that had commenced well before my arrival.

When the students and I viewed the pro-environmental video *In These Ancient Trees,* the tense atmosphere in the classroom increased. Then, upon viewing the Caterpillar-released[2] pro-logging video *Our Continuing Forest,* five students stood on their chairs and applauded. When asked about this behaviour, Eric, a member of the cheering section, offered only the somewhat pleading: "loggers are people, too."

At the beginning of the class period (before any discussion or video viewing), each of the students had been given a sheet of paper with "loggers" typed at the top left-hand side and "environmentalists" typed at the top right-hand side. All those present were asked to write down the first words or phrases that came to their minds when they read each term. The results were startling; four sets of responses are presented here.

	Loggers	Environmentalists
Kristin: (age 13)	dumb, worthless, uncaring	cool, caring, trees, water
Michael: (age 13)	bad, low paying, high-school dropout, bums	smart, mostly girls, wimps, Harvard grads
Paul: (age 12)	hate spotted owls, want to keep their jobs, cut down trees, hate environmentalists	like spotted owls, like environment, save trees, hate loggers
Katrina: (age 13)	people, cutting trees down, giving people wood for their house	people against loggers, put loggers out of work, letting trees fall on their own without supervision so they have a better chance of killing someone, "tree lovers"

"Dumb," "worthless," "cool," "caring," "hate environmentalists," "hate loggers." The students' piercing yet unaffected replies return me now to the dispute's peak years as quickly as could any memory-soaked smell or sound. Here was an age group likely to echo the opinions of their adult attendants, though not yet in the discursively diplomatic manner typical of this White, working- and middle-class enclave. Here was the raw material, so to speak, of the forest dispute.

Several legal decisions led to this classroom moment, and they did so by paralyzing the Pacific Northwest's economically dominant timber industry. In a set of landmark cases from 1989 through 1993, the federal courts found that the National Forest Management Act and the Endangered Species Act, both of which had provisions to protect species, had been clearly and consistently violated by the timber industry. The courts ruled that public forests were not being logged in a sustainable fashion and that no suitable policy had been developed to protect the endangered northern spotted owl. The sale of any new timber on Forest Service and Bureau of Land Management lands west of the Cascade Mountains in the Pacific Northwest was to cease until a legally credible and scientifically sound plan could be enacted.

To those directly employed by the wood products industry the decisions were harsh and abrupt – the result of an overzealous, ill-informed tide of sympathy for the natural world. To environmentalists, they were the culmination of decades of pressure on government and industry, who were perceived as having a destructive and all-powerful hold on the fate of Pacific

Northwest forests (Dietrich 1992). Throughout this period loggers organized a series of protests, one of which resulted in logging trucks blocking traffic on Oregon's main interstate for much of an afternoon (Durbin 1996). Environmentalists countered with equally vociferous gatherings, concerts, and so forth. Bumper stickers appeared everywhere, ranging from: "I Love Spotted Owls ... Fried" to "Stumps Don't Lie." Environmentalists saw the need to ensure the survival of the spotted owl as a way of saving the last stands of old growth remaining on public lands, whereas loggers saw it as a way of disrupting a culture and livelihood whose existence depended upon logging. Each side accused the other of destroying a way of life, of being intrinsically evil, of being insensitive to the needs of the human population (present or future), and of having no appreciation for the natural world.[3]

In short, the Pacific Northwest erupted into a fierce territorial and cultural conflict. Each side was bent on controlling not just common public forestlands, but also the political as well as the intellectual and cultural authority that enabled them to determine just how we ought to care for the region's remaining old-growth temperate rain forests.

Culture and Power

Since the 1960s, anthropologists have defined culture as consisting of shared "webs of meanings," "moral outlooks," and/or "worldviews" that are internalized by and reflected in the behaviour of its members (Geertz 1973; see also Ortner 1984). More recently, this definition has been challenged as unnecessarily essentialist (Clifford 1988, 1997) and/or as ignorant of specific situational contexts as well as caste- or gender-based positional variation (Bourdieu 1993; Holland et al. 1998; Ortner 1999; Strauss and Quinn 1997). Culture, these contemporary theorists have argued, is not a giant symbolic or structural mechanism that imprints itself on the individual and so directs behaviour; rather, it is an overarching, multi-origined, and multi-faceted resource. Individuals draw upon this resource while manipulating it to fit both their own ends as well as the context and social positions from which they act.

Furthermore, this contextually fluctuating use of cultural resources (meanings, symbols, and the restraints imposed by such structural features as race, class, gender, legal and political institutions, etc.) manifests itself in everyday group identities (e.g., as Earth Firsters, anti-choice activists, pro-choice activists, Wall Street brokers, recovering alcoholics, etc.). These identities do not arrive on the socio-political landscape fully formed; rather, "[they] are improvised – in the flow of activity within specific social situations – from the cultural resources at hand" (Holland et al. 1998, 4). Culture is constraining in that it puts boundaries around the range of possible actions; yet it is also flexible in that those boundaries are permeable. Small, seemingly insignificant behaviours, as well as large ideological struggles, will find the weaknesses in any social form as people – agents of cultural change – craft various strategic ways of being, some of which become the basis for new, or at least radically altered, cultural systems.

The relationship between culture and the emergence of identity can be illustrated via two contradictory observations pertinent to the forest dispute: The first is that loggers and environmentalists seemed to talk past one another; each would talk about politics, science, and the forest as though the other didn't exist. For instance, a logger would tell me that no timber had been sold in the Forest Service district in which he (and it was nearly always "he") worked since the court-generated logging injunctions. He would say that it was very difficult to find work in the woods, that preservationists had shut down the forests. An environmentalist would later point to the same district and tell me that several million board feet had been cut from that particular forest. Referring to the years preceding the legal injunctions, he or she would say that these forests had been overcut and that even a complete halt to logging would be insufficient. Paradoxically, both would be right, and both had managed to ignore (talk past) the other's charge. At the same time each party sounded oddly similar, as though each were notably aware of the other's positions.[4] Thus, both timber advocates and environmentalists talked about the joy of being close to nature, about the practice of forest science, about being the *real* victims of larger economic and political forces, about the implication of past cultures for future land use, and about being emotional activists.

Comprehending this discordant "talked-past-yet-sounded-similar" impression requires that one attend simultaneously to (1) dominant cultural discourses about nature, which influenced much of what both loggers and environmentalists had to say; (2) the interplay of competing activist discourses (i.e., the interplay of subordinate discourses and the disparate meaning systems reflected therein); and (3) the fact that both of these discursive layers are ongoing and intertwined at all times. Aggregating these three points, there existed in the Oregon context a dominant though multifaceted discourse that situated both how nature (especially the local forests) should be defined and who would be allowed, politically speaking, to voice that definition. By dominant discourse (an oft-used tag of late) I refer to what Scott (1990) terms "official transcripts" and part of what anthropologists call cultural systems. These are systems of thought that include entrenched conventions for behaviour as well as official definitions of "reality" asserted by the powerful. In this case, the powerful includes but is not limited to scientific, congressional, economic, or corporate bodies whose force "derives in part from their ability to impose [their] construction[s] of reality as the natural order of things" (Philibert 1990, 266). Thus, the scientific community had aptly convinced the federal court that owl habitat was imperiled and that science was central to arguments about future forest management.

Political and regulatory officials had determined, coterminously, that forest management plans would respect the perspectives of local grassroots stakeholders even as officials decried the political "problem" of increasingly "irrational" activist groups. Industry officials meanwhile asserted that life in Oregon as it was known would perish and the economy collapse if logging ceased.

Identities – in this case the competing self-definitions of activist groups (i.e., loggers and environmentalists) – emerged from within this larger cultural frame as each group manipulated its references so as to reflect features of the dominant system in such a way that they shaped its particular vision of a new and better world. Their acts are what Scott (1990, xii) refers to as the "fugitive political conduct of subordinate groups" and what identity and new social movement theorists refer to as the creative means through which human agents both consume and reconfigure cultural systems (Holland et al. 1998; Holland and Kempton 1999). It is for this reason, for example, that environmentalists and loggers reflected the importance of the dominant discourse of science and scientific opinion on owl habitat and forest ecology by speaking frequently of science. Yet – and herein lies the identity-constructing "fugitive" quality – each group managed to put forth very different conceptions of science (i.e., to talk past or to ignore each other on the subject of science). Environmentalists referred to science in its abstract mode, while loggers referred to it in its applied mode. Abstract science worked for environmentalists because it was a means of endorsing identities based on a protective relationship to the forest, and of reinforcing their intersensory bond with nature. Timber advocates endorsed an applied agricultural model of forests because such conceptions countered scientific definitions of the forest as fragile, irrecoverable systems and, instead, promoted the forests as "working" rather than as wild or recreational places.

Identity Dialogues

Identities can thus be seen as dually expressive phenomena, as indicative of (and/or marked by) larger cultural forms (e.g., enduring discourses about the importance of science) and as flexible vehicles through which to challenge those forms. Understanding cultural production requires, further, that analytic attention be paid to the innovative and imaginative actions through which this challenge is achieved. Dialogue is basic to these challenges, particularly dialogues of identity. As an interloper progressing through months of fieldwork, I was often viewed as an instrument of exchange. I told those I met that I was working both sides of the dispute, so it was only natural that I was perceived as a harbinger of opinion from the other side. When timber advocates spoke to me, they were implicitly speaking to the environmentalists with whom I also worked, and vice versa. For this reason, I first thought of this dialogue as an artificial consequence of my research activities.

Eventually, however, it became apparent that numerous scholars have confirmed the ubiquity of contentious dialogue and its connection to identity formation (Bahktin 1981; Holland et al. 1998; Holland and Lave 2001; Johnston, Laraña, and Gusfield 1994; Mead 1934; Taylor 1992). This book draws on their work, focusing especially on identities as negotiated and situationally constructed through dialogues of difference – dialogues aimed ultimately at rewriting the cultural landscape. It paints a fluctuating portrait of competing activists' oppositional dialogue, as well as the changing definitions of culture and nature advanced by disputants. It also demonstrates that, for two reasons, oppositional dialogues are basic to identity formation. First, activists are, by definition, concerned with altering the status quo by stating grievances (Johnston, Laraña, and Gusfield 1994) and imagining new and better worlds. "By modelling possibilities, imaginary worlds can inspire new actions," some of which become new cultural worlds (Holland et al. 1998, 49-51). One way of invoking identity is by making repeated public statements about who one is and how different one is from one's opponent. Indeed, new social movements often come into being because a group's identity is regarded as threatened (Johnston, Laraña, and Gusfield 1994, 23). The second reason for the centrality of oppositional dialogues is that mobilization is, in large part, determined by the staking-out of identity-centred territory. Identity has become an essential ideological tool through which people become collective actors. Invocations of identity are used to call forth "a powerful sense of common cause against those striving to impose a [different standard of] ... personhood and vision of collective life" (Rouse 1995, 23). It is a means of taking action against those who stand in the way of an imagined, better world.

Identity construction is also tied to people's ability to be innovative and agentive. It furthers their ability to respond to situations and overarching cultural contexts by "opportunistically using whatever is at hand to affect their position" (Holland et al. 1998, 279). Impromptu actions are used at those moments where existing cultural resources do not fully meet the requirements of collective actors, thus compelling one to improvise by using the symbolic and linguistic tools at hand in order to craft new possibilities (17-18). So, as is shown in Chapter 4, timber advocates realize that the status of environmentalists as grassroots activists is relatively secure, while theirs is not. Loggers consequently employ the myth of Paul Bunyan, the language of stigmatization, and the symbol of "black-hat" cowboys to shore up their tenuous status as activists and to present themselves as a discrete social group worthy of public sympathy.

All battles about the physical environment also come down to battles about place (whether real or imagined) and the ties between place and identity (Gupta and Ferguson 1992). The places that inform this book are the last stands of old-growth forests in the western Cascade Mountains –

stands that once stretched uninterrupted for hundreds of kilometres from British Columbia to northern California. Durkheim (1965) once noted that moral communities, like churches, function to provide a sense of identity and psychic rootedness. More recently, Basso (1996, 143) draws on Heidegger when speaking about wisdom, place, and physical rootedness. He reminds us that "sensing place is a form of cultural activity." Through this sensing of place "men and women become sharply aware of the complex attachments that link them to features of the physical world ... Places possess a marked capacity for ... inspiring thoughts about who one presently is, or memories of who one used to be, or musing on who one might become" (106-7).

In this sense, both environmental activists and timber activists can be said to make up communities attached to places. The mobility of populations and mass communication have meant that very few communities (or, to borrow Lave and Wegner's term, "communities of practice") are actually integrated, geographically bounded wholes; rather, they are made up of people in separate places (e.g., whether environmentalists in Oregon's cities, or loggers in its rural settings) effectively becoming a single community through the continuous circulation of people, money, goods, and information (Lave and Wegner 1991; Rouse 1991). Matthew Carroll (1995) defines the logging community as a set of identity-based networks rather than as residents of territorially specific locales. Geography is a constant only in that the activists discussed in this book all live in western Oregon and all share an affinity for inspirational places that has informed their respective biographies.

Outline of Chapters

One need not read this book in a linear order (although I'm assuming that it will be so read); one need only accept its basic questions, which are: (1) In what sense can we be talking about a cultural battle when the terms of the debate appear to be driven solely by different legal, scientific, and land-management disputes? and (2) How do competing activists (loggers and environmentalists) operate as cultural producers who reflect and contest these formal terms of the debate?

If you are unaware of the social, historical, and scientific definitions of Oregon's forests, then a close reading of Chapters 2 and 3 will help. If you are concerned primarily with activists' competitive engaged dialogues, then you need only peruse those that most interest you, be they dialogues about emotional meanings (Chapter 7), cultural authenticity (Chapter 6), science (Chapter 5), or grassroots legitimacy (Chapter 4).

That said, the specific sequential logic of the chapters is as follows: Chapters 2 and 3 provide the basis for an informed ethnographic and cultural analysis of the old-growth debate in the Pacific Northwest. Chapter 2 compares current definitions of old growth with those that were popular in earlier times. It includes a brief natural, social, and intellectual history of these forests. This is followed by a look at the parallel ethical legacies that emerged during the formation of public lands policy. What we now call a conservation ethic was embodied by Theodore Roosevelt's colleague, Gifford Pinchot. Pinchot understood that forests (and, therefore, timber supply) were finite and, thus, in need of protection (Hirt 1994; Robbins 1997). He also believed that forests were crops that could be harvested and replanted indefinitely. The land ethic was endorsed by John Muir and Aldo Leopold, and it emphasized the environment's spiritual, aesthetic, and systemic qualities (Shabecoff 1993). Broadly conceived, both ethics express a concern for nature. This explains why both loggers and ancient-forest activists think of themselves as invested in and protective of the forest.

Chapter 3 offers a portrait of the emergence of two grassroots movements in the American west basic to the forest dispute: the Ancient Forest Movement (i.e., environmentalists) and the Forest Community Movement (i.e., loggers). It also offers an ethnographic look at the actions in the recent past of the everyday worlds of cutters, loggers, and front-line environmentalists. Two starkly different expressions of work in the woods are revealed. One finds that both loggers and environmentalists insist upon their concern for nature. For loggers, this is manifested in their pleasure with forest regrowth, their delight with the experience of being able to spend their working day in the woods, and their humility before the force embodied by a falling tree. For environmentalists, this is manifested in their deep preoccupation with the spiritual, aesthetic, and material complexity of old growth as well as their willingness to commit their minds and bodies to protecting it. Also included in this chapter are some of my ruminations concerning the tense circumstances under which I conducted my research: the uneasy shuffle between opponents that was made all the more dissonant by my unvoiced reflections.

Chapter 4 addresses the competition between activists for grassroots status. Who has the best activist credentials, timber advocates or environmentalists? For timber advocates, the construction of identity-centred social groups is connected to their perceived stigmatization. Stigma and identity coalesce as part of a dynamic struggle to achieve political legitimacy and public sympathy. Timber community advocates summon their experiences of stigmatization in order to create politically effective group alliances. They invoke their experiences as victims in order to form a logging culture worthy of protection. Environmentalists' status as political activists was comparatively intact; consequently, they tended to focus on counteracting constructions of loggers as culturally unique by referring to them as ineffectual pawns of the forest industry and by referring to timber communities as pathological. This reinforced the image of environmentalists as the premier grassroots activists and, in so doing, dismissed the groundswell of a logging community movement in the American west (Brick 1995). In the end, the implicit debate over who is pawn and who is victim is a persistent tug-of-war, an ongoing negotiation about legitimacy between two parties struggling to promote their respective goals.

Chapter 4 also introduces the implications of identity-driven social actors competing with one another for public support. Of particular concern is how studies of identity, agency, and culture seem to be preoccupied with single marginalized groups or social movements. This preoccupation ignores the basic fact of environmental conflicts in much of North America. Most battles involve competition between natural resource workers (loggers, ranchers, fishers, miners, etc.) and what Bron Taylor (1995) refers to as ecological resistance movements. Each side is as concerned with the other as it is with the state, cultural forces, or legal-political constraints. This, in turn, means that, if either social movement is to be successful (i.e., to reap some benefit from invoking its identity), then it must come across both as subordinate and as superordinate to its opponent (i.e., as both truly grassroots and as wiser, better, and more powerful). And it must do this at one and the same time.

Chapter 5, as noted above, discusses differing conceptions of science – a subject that, at first glance, may seem unrelated to questions of cultural identity. It begins by noting then president Clinton's official endorsement of a science-based solution to the forest dispute and thereafter examines the troubled implications of this endorsement. It demonstrates that two very different notions of science are at the centre of the forest dispute. Loggers prefer an applied science based upon common-sense empiricism of the "seeing-is-believing" variety, while environmentalists prefer abstract science and the beauty associated with complexity and holistic systems.

Chapters 6 and 7 depart from their predecessors to the extent that they take up more fully the roles of agency and imagination as they apply to the reconfiguration of cultural forms. Chapter 6 finds that both loggers and environmentalists recognize the significance of cultural history and authority to the dispute. This recognition is manifested, primarily, in activists' deference to the symbolic power of Aboriginal land-use traditions. This (symbolic) power is rooted in mainstream American notions that "Indians represent the possibility that there are individuals who are 'naturally' born into a way of life that effortlessly embodies principles of Western conservation" (Conklin 1997, 722). Both environmental and timber activists recognize that their authority regarding past and future land use depends on the ease with which they can play into publicly salient ideas about past peoples as ecologically instructive due to their relationship with physical territories. Practically speaking, this means that activists affiliated with the more "Aboriginal," or "authentic," tradition wield a distinct political advantage over those who do not. However, the question of cultural legitimacy presents a problem for both activist parties because both groups have markedly equivocal claims to authentic status. Environmentalists can easily be dismissed as very few live outside urban areas, and few extract a living through physical labour in the natural world. One way to gain advantage under these circumstances is to ally one's group with those who already possess authenticity and, in so doing, acquire, albeit vicariously, the cultural capital that authenticity provides. Discussions about past peoples, like discussions about grassroots legitimacy, are used to bolster one's own fledgling cultural legitimacy by sheer force of association.

Loggers, conversely, appear uneasy with the strategic advantage embodied in the ability of environmentalists to capitalize on mainstream American notions of Aboriginal ecological nobility. When addressing the behaviour of their opponents, loggers seek to redefine popular images of an Aboriginal past while simultaneously recasting their own group as analogous to Aboriginal peoples and, thus, as deserving of authentic status and its concomitant rights. Loggers, too, have their own vision of future land use, which they regard as rooted in a recent past – one that improved upon Aboriginal practices. They regard any attempts to mimic Aboriginal traditions as introducing (not alleviating) ecological danger, and they equate Aboriginal-inspired ideas with a naive and non-natural disruption of the practices and communities of labourers that carry out those practices.

Chapter 7 examines how timber advocates and environmentalists explain the emotional character of the forest dispute. Both groups must contend with accusations of excessive emotionality, with criticisms from the centre (i.e., the status quo) about the appropriateness of their behaviour. Of course, to be an activist (whether on the right or on the left) is to be on the margins of the social whole. Feminist scholars have argued that one method of criticizing or delegitimizing those on the margins – be they labourers, women, people of colour, or the poor – has been to accuse them of excessive emotionality. This labels the disenfranchised as out of control, as being over the imagined line of reasonableness, and thus serves to silence and disempower them (see Lutz 1988). Loggers address this criticism by disassociating themselves from expressions of emotion. They appear forever conscious about being seen as irrational players in the dispute and work hard to counter these impressions in the public mind. Most environmentalists, on the other hand, do not betray any need to control their emotions and, instead, invoke emotionality to extend their definition of community as being comprised of both human and biotic subjects. Thereafter, the communicative power of emotional language is explored by demonstrating how emotion and ethical practice relate to agency, gender, and social class. Close scrutiny of activist invocations of fear demonstrates that both activist groups present fear in moral terms. Loggers equate fear with undergoing the danger of performing physically hazardous work to produce goods for an unappreciative public; environmentalists equate fear with the failure to commit to social change, and with the failure to fully embody an intersubjective bond that they regard as central to the human-nature interface.

Chapter 8, the conclusion, pulls together the strings of evidence for the changing shape of culture both within the academy and across activist groups. It explores the benefits of studying environmental activism as a triangular process that incorporates both the influence of entrenched cultural forms and discourses and the oppositional dialogue between identity groups that work to promote change. I then draw some implications from this study for policy purposes, emphasizing those that concern the move towards public participation in local decisions about land management and the corresponding investigative methods used by policy analysts to ascertain how the public values nature.

2

The Cycle of History: Public Lands, Forest Health, and Activist Histories in the American West

To engage in forest activism is to enter into a debate about the relationship between humans, nature, resource utilization, and ecological survival. Loggers see themselves as having successfully aided nature by taking an ageing, rotting forest and replacing it with a growing one. "When the housed say too much wood was cut, loggers perceive hypocrisy and betrayal" (Dietrich 1992, 26). Among most environmentalists, old growth inspires a kind of worship and a profoundly protective stance towards the ecosystems that precariously house a multiplicity of endangered species. "Above all, the forest is a remnant of the world as it was before man appeared, as it was when water was fit to drink and air was fit to breathe" (Caufield 1990, 46).[1]

Comprehending the conflict between timber advocates and environmentalists invites a related set of foundational questions: What is an old-growth conifer forest and how much of that forest is left? How did we get here in the first place? What are the historical patterns of land use in the Pacific Northwest? Has logging always devastated the forests? Is our concern simply nostalgia for another time or is it materially defensible? Where can we locate the intellectual and activist roots of both sides of the dispute? And, finally, to what policy level has this legacy taken us? Brief answers to these questions provide the stage upon which all subsequent chapters are set.

What Is Old Growth?

A universally acceptable definition of Pacific Northwest old-growth forests continues to escape the scientific community. Materially, the attributes agreed upon by the Forest Service's Old-Growth Definition Task Group will suffice.

- two or more tree species with a wide range of ages and tree sizes
- six to eight Douglas fir or other coniferous trees per acre (.405 hectares), at least 30 inches (76 centimetres) in diameter or at least 200 years old
- a multilayered forest canopy

- two to four snags (standing dead trees) per acre at least 20 inches (51 centimetres) in diameter and at least 15 feet (4.5 metres) tall
- at least 10 tons per acre of fallen logs (deadfall), including at least two sections of fallen logs per acre that are at least 24 inches in diameter and 50 feet (15.2 metres) long (Durbin and Koberstein 1990, 25; see also Franklin and Waring 1980).[2]

Traditionally, foresters defined treed vegetational areas in terms of their climax species: the tree species expected to predominate if growth were to continue undisturbed for centuries. Much of the Pacific Northwest is classified by the potential prevalence of hemlock, a shade-tolerant species that grows well in dense, canopied forests. But the preponderance of Douglas fir in the forests tells another story: more often than not fire catastrophically disrupted the "climactic" spreading of hemlock, making way for the relatively long-lived Douglas fir. Disturbances like fire break up dense forest canopies, eliminate crowded underbrush, and thus favour the sunshine-thriving Douglas fir (Booth 1994, 30; Pyne 1997).

Pacific Northwest forests contain more living plant matter (biomass) per acre than does any other studied forest on Earth (Waring and Franklin 1979, 1,380). The millions of needles on a mature Douglas fir (70,000,000 is not uncommon) act as a kind of comb that sweeps drifting clouds and fog for moisture (Dietrich 1992, 104). One would think that the canopied density of an old-growth forest would block the rainfall that feeds watersheds from reaching the forest floor. On the contrary, however, about 35 percent of the water in an old-growth conifer forest is the product of fog drip. When only 25 percent of two drainages in a Portland-area watershed was logged, stream flow from within the logged drainages decreased despite the conventional wisdom that anticipated an increased flow of water (Caufield 1990; Maser 1989; Norse 1990, 147).

Snags (standing but dead trees) and deadfall – rotted, decadent wood to most loggers – are the basis of much of an old-growth forest's health and biological diversity. They are the dwelling place of hundreds of vertebrate and thousands of invertebrate species. The decaying material is recycled into the soil and the forest. The dead trees are essential to forest health, the basis of its astounding productivity because much of the forests' nutrients reside in the living and dead plant material itself and not, as one might expect, in the soil. "One-third of the soil's organic matter comes from decaying logs" (Caufield 1990, 49-50). Clear-cut logging is criticized as indefensible because it denudes the forest of this organic legacy – a legacy upon which the next biotic generation depends (Durbin 1996).

The importance of the northern spotted owl lies in it being an "indicator species." In lay terms, the owl is the canary in the coal mine: just as the demise of the canary draws attention to the toxicity of the mine, so the demise of the owl draws attention to the ill-health of the forest. There is, borrowing heavily from Catherine Caufield's (1990) dense and comprehensive article, "The Ancient Forest," an intimate connection between the owl's survival, the forest's decay cycle, and certain fungi. Mycorrhizal fungi are of particular importance as they grow on decaying trees and on the roots of living trees. These fungi infect the roots of many tree species, including most of the conifers in the Pacific Northwest forest. In doing so, they promote the growth of tiny root hairs that spread across the forest floor searching for nutrients that are unavailable to uninfected roots. Without mycorrhizae, trees cannot obtain the phosphorus, the nitrogen, and the water they need to survive and grow (50).

Caufield documents attempts by experimental foresters to get seedlings to grow without the benefit of these fungi: all of the saplings died within two years of planting. Small mammals such as mice, squirrels, chipmunks, and voles, who consume the fungi and then distribute them via their scat, are responsible for spreading fungi spores throughout the forest. These fungi and truffles must come into contact with tree roots if the latter are to benefit (Caufield 1990; Maser 1989). A reduced spotted owl population quite likely means that there are not enough truffles and fungi to support the small mammals that are their prey. In other words, a low owl population assumes a low small mammal population and, therefore, a limited dispersal of the mycorrhizal fungi upon which tree growth depends.

Lay Perceptions
William Dietrich (1992, 23), in his superb journalistic (and occasionally ethnographic) account of the forest dispute, writes of his inability to comprehend "how so many good people could love the forest so fiercely in such completely different ways." Some, for example, are captivated by the Pacific Northwest conifer forest's ability to grow among the finest-grained woods in the world, while others are captivated by its genetic complexity and spiritual power. Both these attitudes are reflected in the social and natural history of the region.

Most loggers and representatives of the timber industry look at these forests and imagine the financial and utilitarian benefit derived from harvesting wood that will soon decay. In its wake, they envision a new, neater forest consisting mostly of the more desirable Douglas fir. Timber cruisers (those employed to evaluate the output of a potential timber sale) are respected for their talent at "guesstimating" the number of board feet (1 foot by 1 foot by 1 inch) in a particularly large tree. The average single-family home requires 10,000 board feet of lumber, and one very large Douglas fir can contain as many as 30,000 board feet (Booth 1994; Caufield 1990, 58).

Loggers quite often refer to old-growth forests as "decadent." At first glance, given its association with Dionysian (usually urban) indulgences, this is a rather strange adjective to apply to forests. But this term does epitomize a critique common to many who oppose further protection of old-growth forests. First and foremost for these critics is the belief that it is utterly wasteful to allow merchantable standing timber to exceed its wood-producing prime by rotting to death.[3]

For timber enthusiasts, the idea, popularized by a Weyerhaeuser ad, that Oregon might one day "grow out of trees" is absurd. Travelling dusty roads with a timber cruiser for a small old-growth mill in Oregon's coast range (an event described in some detail in Chapter 3), I noted his confidence that the next generation of trees will be a newer, stronger breed consisting of the most desirable species. He fussed like an attentive master gardener over the young forest on either side of the road we travelled. The trees were bright green, the by-product of cutting and replanting that took place thirty years ago. To me, the sight of these even-aged stands was monotonous, relatively unimpressive; but to my guide they were cause for great concern. He was upset by breakage caused by heavy snowfall, even though the growth set-back will be irrelevant to his own arc of employment. The trees are supposed to be cut according to a 100-year rotation: they will be harvested long after he is dead.

Environmentalists do not generally share the logger's awe of "board-feet" productivity. They talk instead of biomass, biodiversity, aesthetics (old-growth forests are likened to Europe's ancient cathedrals), and spiritual inspiration. Their sense of urgency is rooted in the conviction that to destroy our habitat is to destroy ourselves. Some extend their egalitarian sentiments to all living beings and proclaim a willingness to decrease their material standard of living in order to protect other life forms. Further, for them, human-centred notions of usefulness are not paramount; nature is believed to have worth in and of itself.[4]

In the end, the forest ecologist's delight with the overwhelming complexity and interdependence of forest life is shared by most environmentalists and much of the lay public. Foresters Jerry Franklin and Kathryn Kohm (1999) summarized the last three decades of research on the Pacific Northwest's old-growth conifer forests, referring to them as demonstrating the forests' extreme regulatory and structural complexity, their high level of biological diversity, and their fire and disturbance "legacies," which enrich subsequent (postdisturbance) forest ecosystems. "These and other findings have highlighted the marked contrast between natural forest ecosystems and intensively managed plantations, and between natural disturbances and clearcuts" (243). This complexity is perfectly captured by ecologist Frank Elgin's statement: "Ecosystems may not only be more complex than we think, they may be more complex than we *can* think" (Dietrich 1992, 110).

Altered Landscapes

Thus far, a picture has been painted of contemporary expert and lay perceptions of old-growth forests. But this portrait is falsely static, for it conceals a dynamic history of ideas about nature and about land use as conceived of, and practised by, Aboriginal populations and early White settlers, respectively. It also conceals the story of an expanding twentieth-century population and the industry upon which it depended.

In the Oregon neighbourhood in which I lived when I started writing this book, there were flyers stapled to telephone poles announcing the availability of training in Native American (no particular nation or band was specified) "shamanistic" and "survival" skills – skills that would enable one to live in harmony with nature. The teacher was not himself Native American.[5] This poster, without betraying awareness of the fact that different Native North American groups had different ways of living on the land, reflects the romantic, decontextualized borrowings from Aboriginal lore that pervade some spheres of the environmental community – a phenomenon carefully examined in Chapter 6.

Ronald Gautier (a logger-cum-tree farmer working 1,000 acres [400 hectares] of family lands scattered about the north end of Oregon's Willamette Valley) likes to point, alternatively, to the surrounding forested hills and recall the fact that the Calapooia had intentionally deforested the area with fire. Ronald's message: we (descendants of European colonists) are not the first land managers, nor are we the first to clear the forests. The historical record confirms this. Two million of the Willamette Valley's acres were maintained in prairie and savannah as a consequence of Aboriginal-set fires.[6] After the fires, wild honey, grasshoppers, and tarweed seeds were more readily available for gathering, and a higher concentration of game animals congregated in the cleared areas (Boag 1992; Robbins 1997; White 1980). Citing multiple historical records, Pyne (1997) notes too that early White settlers were comparatively alarmed by the annual firing of Oregon's Willamette and Tualatin valleys to harvest wild wheat and to hunt, although they understood very well the underlying purpose:

> We did not know that the Indians were wont to baptize the whole country with fire at the close of every summer ... [until] the whites prevented them ... The bands ... united in the annual roundup ... At a given signal ... they commenced burning off the whole face of the country and driving wild game to a common center. There was considerable skill required to do this correctly ... the best hunters went inside and shot the game they thought should be killed.

Postsettlement, however, the area quickly returned to its forested cover: "Within a few years ... the hills and prairies had already commenced to grow up with a young growth of firs and oaks" (336).

The impact of Aboriginal practices on forest consumption ought not, however, to be overestimated. Booth (1994, 55) estimates Aboriginal timber consumption (per person, annually) to be about 15 percent of today's consumption figure. To this he adds that salmon, the mainstay of Aboriginal subsistence, were extracted at levels that were much lower than those of today. His arguments about Aboriginal rates of consumption and use of fire are twofold: (1) Aboriginal inhabitants exercised restraint when drawing on the natural world for their needs; and (2) the (pristine) wilderness of our imaginations, the world that preceded European settlement, was undeniably a territory shaped by its inhabitants to fit their own purposes.

To the best of our knowledge, Aboriginal manipulation of the environment occurred, whether due to conscious intent, to the honing of practices over thousands of years, or to population scarcity, without threatening the area's ability to sustain its inhabitants and without the introduction of exogenous biological life. Some species were clearly favoured and fostered over others, but two vastly different biological worlds did not come into contact with one another until there was a substantial European presence on the continent (White 1980, 26-36). Moreover, the decimation of Aboriginal populations via disease, warfare, and colonization, as well as by the government-imposed reservation system, meant the end of methods of land use that were unique to Aboriginal settlement.[7]

Subduing "Wild" Nature

The timber industry's economic domination of the Pacific Northwest was not an immediate result of White settlement. Initially, the timber industry evolved despite government policy that emphasized the agricultural development of the west. The desire to get "the land subdued and wild nature out of it" (so that farming might begin) was what motivated most nineteenth-century settlers (Boag 1992; Booth 1994, 73; Robbins 1988). "The settlers never thought of their axe work as deforestation but as the progress of cultivation" (Shabecoff 1993, 30). The federal government's Preemption Act, 1841, and the Donation Land Laws, 1850, provided anywhere from 160 to 320 acres (64 to 130 hectares) to those who could (and would) become resident citizen farmers (White 1980, 37-38). The legal validation of settler occupation provided by the Oregon Donation Land Laws was a particular inducement to immigration. Before the act's 1855 expiration, 25,000 to 30,000 people of European descent arrived in Oregon Country, representing an "increase in that population of nearly 300 percent" (Robbins 1997, 83).[8] Overall, a pervasive spirit of consumption was matched by the availability of forcibly appropriated Aboriginal land.

Biologically, the influx of transported exogenous crops that accompanied the human settlers vastly simplified the ecological world. The soil that supported the abundant forests was not easily converted to productive farmland. Except for some river valley and prairie areas, the ground was generally dry, sandy, and unsuitable for anything but conifer forests.[9] Domesticated crops perished in the absence of intensive human labour. Nonetheless, federally (and distantly) mandated laws persisted, and legislation blindly promoted the "progress of civilization" by continuing to allot land to tenant farmers. Lawmakers believed that clearing the land of trees increased rather than decreased the land's value. Meanwhile, loggers and mill owners became cognizant of the economic potential of these devalued forests and managed to evade federal laws designed to place land in the hands of tenant farmers. Tolerance for fraud and theft filled the gap between the federal government's agricultural policy (which was extremely anti-tree) and the settlers' localized (and lucrative) logging opportunities (Robbins 1988; Shabecoff 1993; White 1980).

In the 1860s and 1870s mills began to appear in the Pacific Northwest, providing the growing west coast population with lumber for housing and civic infrastructure. As much of the public domain had yet to be surveyed, it was common for independent loggers to provide mills with timber by simply cutting whatever and wherever they could. Those who formerly had cut timber for their own homes began, in the absence of any title or right to the land, to cut whole sections for mill supply (White 1980, 80). During this period punishment for such illegal actions was inconsequential precisely because so few people conceived of the forest as a benefit (Booth 1994, 80-82). "Once begun, however, cutting timber on public lands easily slid into a habit, and even into something of a right ... In 1877, investigators claimed that half the timber cut from Washington State's Puget Sound region had been illegally taken from government or railroad land" (White 1980, 82).[10]

Mill owners and/or lumbermen interested in acquiring their own land so that a regular timber supply could be secured did so by subverting the Preemption Act and the land acquisition laws. Proprietors would hire "dummy entrymen" to go into land offices and pose as potential tenant farmers, whereupon they would take title in their own or their company's name. At the same time, mill owners continued to purchase timber from loggers, saving their own land for anticipated periods of scarcity. Thus a solid tie was established between mill owners and loggers: the former provided the latter with a market, and the latter obtained their equipment from the former.

The Homestead Act, 1863, and the Timber and Stone Act, 1878, reiterated, in theory, the federal government's desire to preserve land for settlers, but mill owners continued to use the laws to expand their holdings. The Puget Mill Company had entire crews from its lumber schooners file false homestead claims. A comparison of actual land holdings with legally transacted purchases found that 38 percent of private forest holdings was acquired through the illegal use of the Homestead Act and similar acts (Booth 1994, 81). It was during this period that one could also acquire land via railroad grants. The government would provide land to railroad companies in exchange for the building of rail systems. In order to finance this construction, railroad companies would sell considerable portions of their granted lands to mill owners.

To summarize, US land distribution policy during the above decades was supposed to result in the agricultural colonization of the west; in actuality, it resulted in considerable acreage being acquired by timber companies. The social and political climate was such that illegal access to timber was the norm; the lumber industry took shape within an atmosphere that tolerated theft, fraud, and bribery.

Ecological Impact of Working the Land

During the late nineteenth century the fledgling timber industry's impact on the forest was marginal (Boag 1992; Robbins 1989, 1997). Timber removal was extremely labour-intensive; it could take days for a logger to remove one or two large spars. Only a few giant trees were extracted from an area at a time. Trees were felled with axes and cut into movable sections with crosscut saws. Sections were hitched to a team of oxen and dragged to the ocean (or river) along a skid road, "a corduroy track heavily greased with dogfish oil" (White 1980, 87). In order to ensure transportation, logging always took place in the vicinity of water, thus only the periphery of the forest was logged. These practices left the forest's integrity and gene pool intact, and they promoted many of today's healthy late-successional forests. Today, the avoidance of large seed- and gene pool-decimating clear-cuts is the basis for the promotion of smaller more frequent clear-cuts and selective logging – the assumption being that intermittent cutting over a large area is less destructive than is clear-cutting/extracting the same number of trees over a small area.[11]

Industrial Logging

Historian and one-time Coos Bay, Oregon, logger William Robbins (1985, 1989, 1997) argues that industrial forestry on the Pacific slope is a twentieth-century phenomenon. A set of technological innovations, a dramatic increase in population, and the availability of large-scale capital propelled the twentieth-century development of the Pacific Northwest. Drastic changes in the structure and biological composition of the forest were the result of the tremendous desire and potential for profit provided by the use of the donkey steam engine and the narrow gauge railroad. The logging railroad gave loggers access to previously untouched stands of timber. Even the early versions of the donkey steam engine substantially reduced the cost of timber harvesting and enabled winter logging (heretofore, oxen would become mired in mud during the long, wet winter). The engine powered first one and later as many as five winches that hauled in the fallen logs. Loggers attached cables from these winches to metal chokers, which were wrapped around the logs. Massive stands could thus be cleared (Booth 1994, 76; Robbins 1988). High-lead logging followed: "The high lead was added to earlier ground-lead logging in about 1910; operators were able to move logs with one end suspended in the air, a technique that greatly increased both the volume of timber that could be moved and the incidence of injury and death to workers" (Robbins 1990, 6).[12]

The extension of the railroad into the Pacific Northwest brought with it both an enormous population boom and considerable capital investment. In the first decade of the twentieth century, the population of Washington State increased by 120 percent: 62 percent in Oregon and 60 percent in California (Robbins 1989, 235). Expanding markets, human migration, technological innovations, and "speculative mania in an investor's frontier" paved the way for an ongoing cyclic pattern. "The big profits were in cutting, stripping and then moving on to the next stand. The dynamics and logic of a social system in which profit and loss were the major criteria for land management decisions both created and impoverished lumber towns" (Robbins 1985, 417).

Companies rallied to obtain as much timberland as possible, particularly since the largest profits resulted from land speculation rather than mill owning. Massive speculation left huge volumes of timberland in the hands of single investors – Frederick Weyerhaeuser being a case in point (Robbins 1988, 27). Moreover, the purchases had to be paid for by cutting rapidly and extensively. This triggered an excess of production over demand. The resultant instability only increased the need to liquidate timber as people desperately attempted to avoid bankruptcy during over-supply periods (Robbins 1989, 236). By the late nineteenth century, regional and national trade organizations emerged in an effort to predict market demand, regulate timber supply, and rationalize practices. But, for the most part, the boom-bust cycles persisted and thus ensured that the timber industry "was not and could not be a humane system" (236). The system's inhumanity also inspired the popularity of radical labour movements, such as the International Workers of the World (Wobblies) (237).

Industrial logging marks the beginning of the forest devastation that can now be seen in aerial photographs (Devall 1993). Many timber workers and representatives have come to resent these photographs, insisting that they show only the worst cases. However, long-time logger and field consultant Jim Stratton agrees that logging "leaves a horrible mess." The gouged land tends to collect water, which encourages the growth of alder and hemlock. Oregon's recent devastating floods and mudslides have also been attributed to the collection of water and the destabilization of soil typical of clear-cut sites.

The clear-cut's dormant fire hazard was reduced by modern slash-burning practices,[13] but its ecological consequences remained. Cleared land brought an excess of berries, fireweed, and mammals who prefer to graze on infant forest vegetation. Because logging eliminated stands of the virgin forest that had once seeded neighbouring cutover lands, adjacent seed trees were less available than they had been historically. Young forests came to border recently cut forests, cutting off yet another seed source. The pattern typifies what is currently referred to as a "fragmented" forest. Earlier bull-team logging left, alternatively, a large number and variety of trees at the logged site – an abundant, diverse seed source crucial to forest regeneration. Thus today's "New Forestry" – a cornerstone of cutting-edge, ecosystem management – is, in part, a return to the messy, bull-team cuts of early logging.

Public Lands, Conservationism, and Environmentalism

A large body of work documents the late-nineteenth-century/early-twentieth-century presence (and popularity) of ideas that parallel those of contemporary environmentalists and conservationists. The idea, for instance, that the Pacific Northwest forest might be depleted just as had been the lesser forests to the east was alive and well in the early part of the twentieth century. Westerners and public foresters largely agreed that the forests were poorly managed (Robbins 1985, 1989). A 1912 report by an Oregon State forester cautioned against the fallacy that these forests were inexhaustible (Robbins 1985, 416). More broadly, calls for sustainable harvests and sustainable communities increased as "the gospel of efficiency" took hold (Hays 1958; see also Hirt 1994). Efficient use of forest resources was touted as the path to civic and economic sustainability, with primary emphasis being placed on the material wastefulness of most logging operations. Conventionally, anything below the first branches of a felled tree was discarded, and the remaining stumps often stood three to six metres above the ground. It was argued that the waste that occurred could have been of benefit (White 1980, 93) and that only better engineering of the forests would ensure the sustainability of annual timber yields (Hays 1998, 337-38). This "efficiency" legacy lacked explicit concern with forest health per se; nonetheless, it is the basis upon which loggers identify themselves as conservationists who have long embraced an environmental ethic. They see themselves, then and now, as sensitive to "sound" (especially non-wasteful) logging practices.[14]

Much earlier, naturalist George Perkins Marsh (1864), a Renaissance personality from Vermont, published the remarkable *Man and Nature*.[15] At a time when many Americans perceived nature as a behemoth to be subdued, Marsh documented the sensitivity of what would today be called an ecosystem. He criticized the (over)cultivated gardens idealized by the Jeffersonian tradition, deeming them an agent of destruction, and he outlined the impact of logging on watersheds, water supply, salmon runs, and flooding (Robbins 1985, 1-2). Clear-cutting in watersheds, he argued, "resulted in droughts, flood, erosion, and unfavorable climactic conditions"; that is, human-generated disasters that he argued were "responsible for the decline of Mediterranean empires" (Nash 2001, 104). He was among the first to link preservation of wild land with economic well-being, and he wisely cautioned his audience neither to err in the manner of Old World civilizations nor to "wait till the slow and sure progress of exact science has taught us a better economy" (Robbins 1985; Shabecoff 1993, 58).

Marsh's discourse was as scientific as that of John Muir, Ralph Waldo Emerson, and Henry David Thoreau was spiritual-cum-moral. These three men are generally associated with transcendentalist philosophy, which, in its New England form, contemplated nature's capacity for spiritual healing.[16] Nature was regarded as the "proper source of religion." In this sense transcendentalists carried on the tradition of such Romantic poets as Wordsworth, who believed in the "moral impulses emanating from the fields and woods." "In Wilderness," wrote Emerson, "I find something more dear and connate ... [I]n the woods we return to reason and faith." Further, the transcendentalist belief in the moral goodness of nature and the human spirit's ability to thrive in the wilderness contradicted both the frontier era's quest to subdue the land and the Puritan belief that humanity's inherent sinfulness "ran rampant in the moral vacuum that was wilderness" (Nash 2001, 86).

Thoreau and Muir are also closely associated with early efforts at wilderness preservation. Muir was one of the Sierra Club's founding members, and, as will shortly become evident, he inspired many contemporary environmentalists, playing a central role in getting land designated to a national system of forest reserves and parks. Emerson, more theorist than practitioner, lectured on the need for a harmonious synthesis of reverence for nature and tolerance of technology. He believed that nature could recover from the damaging impact of society and technology. Thoreau, meanwhile, was legendary for living in nature and for his fear of technology's impact on the land. He saw wilderness as the counterbalance to the heavy burdens placed on the human soul by labour and the stress of living in an increasingly materialistic, urbanized society (Cronon 1996; Shabecoff 1993).

Though often ignored in historical accounts of American environmentalism, several prominent women (mostly White) writers of the period heightened public sensitivity to the importance of cultivation based on indigenous plant species, emphasized the dynamism of natural systems, and called for the preservation of natural landscapes (Norwood 1993). In 1850, Susan Fenimore Cooper (daughter of James) published *Rural Hours*. Her writing admonished American (particularly female) gardeners for their preoccupation with imported agricultural and garden species and "hothouse plants secluded in an artificial environment" (Cooper 1850). She further implored her readers to "resist ambitious manipulation of God's [natural] creation" (Norwood 1993, 40). On the other hand, Mary Treat, published naturalist and correspondent of Charles Darwin, "viewed nature as much less static than did Cooper" and, given Darwin's insights, "argued against human supremacy in an hierarchical natural world" (41). Towards the turn of the century, Mary Austin, inspired by John Muir, worked on political campaigns to conserve water and wilderness, and she wrote several popular books (e.g., *The Land of Little Rain,* and *The Land of Journey's Ending*), each credited for its influence on American environmental values (49).

Public Lands
The national forest and national park systems were established following the removal of Aboriginal populations to reservation land in decades prior and through the General Land Revision Act, 1891 (Robbins 1985; White 1991). The act gave the president authority to create forest reserves by proclamation. Further amendments to the act in 1897 specified the need to manage forests in order to protect watersheds and furnish a continuous supply of timber to the citizens of the United States. It was assumed that when private timber was depleted, public forestland could be harvested. The act also gave the secretary of the interior the power to regulate and manage reserve lands (Caufield 1990, 1; White 1991, 405-407; Wolf 1993, 1-3).

President Harrison placed three million acres (1.2 million hectares) in six forest reserves in 1891 and 1892, and he added nine more reserves to make a total of thirteen million acres (5.6 million hectares) before leaving office (Robbins 1982, 23). By 1894 there were seventeen reserves with nearly eighteen million acres (7.3 million hectares), and in 1897 twelve new reserves (totalling thirty-nine million acres or 15.8 million hectares) were added to them (Wolf 1993, 2). Early-twentieth-century suspicions of a timber famine on privately owned lands were common, and this motivated the assignation of additional lands to the reserve system (Robbins 1988).

Public response to the forest reserve system was mixed. Westerners criticized the reserves because their existence infringed upon local interests and the rights of small-time settlers. Many dismissed the program as "an eastern conspiracy to keep public lands from the common citizen" (Robbins 1985, 8). Major timber companies supported the nationalization of large tracts of land in the west because it removed acreage and, therefore, competition from the hands of small milling operations and independent ("gyppo") loggers.Dissent, particularly from western senators, continued until, in 1897, President McKinley signed into law an amendment that "opened up the resources in the reserves to managed public use" (Robbins 1982, 26). By

1907, with considerable help from conservationist president Theodore Roosevelt, the system had increased to 159 reserves containing nearly 151 million acres (61.2 million hectares). Twenty-one new national forests were added just before Congress stripped Roosevelt of the power to create reserves by proclamation (White 1991, 407-09).

The initial congressional events that secured and increased the acreage of national forest reserves are fondly connected in the public mind with the presidency of Theodore Roosevelt – a remembrance that harbours some truth and some fiction. Roosevelt was not yet president when the 1891 act that initiated the forest reserve system was established. But he clearly believed in protecting public land from wasteful corporate consumption, and he believed that public land needed to benefit all citizens. These ideals initiated decades of public scrutiny of, and challenges to, Forest Service policy. Of Roosevelt's legacy, Shabecoff (1993, 59-69) writes: "The democratic principle had been established and would never be surrendered by those who cared about the land." Because of Roosevelt, exploitation of the nation's resources would always have to be "justified under the guise of spurring economic growth, protecting jobs, safeguarding national security, or some other subterfuge."

Roosevelt endorsed the ecosystemic ideas of George Perkins Marsh and possessed some of John Muir's spiritual appreciation of nature; however, ultimately (to the aggravation of more than a few contemporary environmentalists), he promoted the economic benefits of forests and placed his trust in the science of forestry advocated by Gifford Pinchot (Robbins 1982). Pinchot was the first head of the United States Forest Service (the first agency to actively manage federal forestlands). By all accounts Pinchot was a charismatic populist, and he was responsible for modern forestry's attention to wood-product output. He was also responsible for the expansion of the Forest Service bureaucracy and the promulgation of the forests as a good for public use (Hirt 1994; Robbins 1985).

To many, he continues to embody the "right" approach to forestry. Trained in Germany, Pinchot imported European ideas of scientific management, holding that trees were a manageable crop (Caufield 1990, 57; Shabecoff 1993, 65; White 1991, 407). Pinchot, like Roosevelt and even Muir, deplored the timber industry's inefficient use of land, which he saw as squandering the forest's economic potential. Lumbermen, he noted, were "concerned with squeezing the last penny from the woods without regards to consequences," whereas the "forester managed them scientifically so as to obtain a steady and continuing supply of valuable products" (Nash 2001, 134). Pinchot also successfully added the national forests to his managerial domain, using the farming motif to justify the transfer of all Forest Service reserve land from the Department of the Interior to the Department of

Agriculture (where it resides today). Aesthetics, recreation, and wildlife protection meant little to Pinchot (Shabecoff 1993, 69); his favourite aphorism was "wilderness is waste" (White 1991, 409). At the same time, the democratic essence of the Pinchot/Roosevelt legacy continues to be revered: Pinchot and Roosevelt were "inventors of the commons," and they promoted a public land base whose purpose was to serve the public good (Worster 1993, 107).

The Split

The rift between Muir-via-transcendentalism-style environmentalism and Pinchot-style conservationism can be traced to (1) the development of policy to manage these new public forests and (2) a subsequent battle over a steep valley within the reserved wild land – land from which Yosemite National Park was derived. Initially, both conservationists and environmentalists fought for the expansion of the forest reserve system. Muir supported Pinchot-style forestry because he believed that anything was an improvement over unregulated timber cutting and because he assumed that some portion of the federally owned forestland would be committed to wilderness. Forests, declared Muir (sounding very like Pinchot), "must be not only preserved but used ... [they must] be made to yield a sure harvest of timber, while at the same time all their far reaching [aesthetic and spiritual] uses [must] be maintained unimpaired" (Nash 2001, 134-35). Indeed, Muir and Pinchot were initially close friends and were mutually involved in the 1895 Forest Congress convened by the National Academy of Science to develop management guidelines for national forests. Yet, by 1897, it became irrevocably clear that Muir imagined the forests as a pool from which to set aside significant undeveloped area, whereas Pinchot imagined "opening all reserves to carefully managed economic development" (136). When Congress finally passed the 1897 Forest Management Act, Pinchot's vision prevailed. Wilderness was not to be the reserves' primary use. Again, as noted above, the act made it clear that the forests' primary purpose was "to furnish a continuous supply of timber for the use and necessities of the United States" (137).

The final dissolution of the peaceful coexistence of utilitarian conservationism and wilderness advocacy occurred during the (in)famous dispute over Yosemite National Park's Hetch Hetchy Valley. Yosemite was "the first tract of wild land set aside by an Act of Congress in 1864" (Spirn 1996).[17] Hetch Hetchy Valley (known for a beauty equal to that of Yosemite Valley) was included in the park's borders under the National Park Act, 1890, following a campaign led by Muir (Long 1995). Over successive years (from 1880 through 1907), city engineers in San Francisco had proposed damming the valley's Tuolumne River in order to establish a public utility that would supply inexpensive power to San Francisco. The battle for and against was fought on ethical grounds, pitting the public good (to wit, wresting the control of power resources away from "greedy" developers) against the spiritual-aesthetic good of preserving a pristine nature. True to their respective populist and spiritualist bents, Pinchot supported the project while Muir vehemently opposed it (labelling Pinchot's supporters "temple destroyers"). Both men solicited President Roosevelt's backing, particularly as the public demand for inexpensive water and power heightened following the 1906 San Francisco earthquake and fires. After considerable indecisiveness, Roosevelt came down on Pinchot's side, and, in 1913, the dam became a reality (Shabecoff 1993, 73-75; White 1990, 412-15).

Leopold's Land Ethic
In the years that preceded the Second World War the demands for timber harvesting on the public lands of the Pacific Northwest were relatively slight. At lower elevations, private timberland was still being cleared of its most lucrative trees; indeed, the overwhelming majority (95 percent) of wood products produced nationally was derived from private land (Hirt 1994). To a certain extent, this allowed an alternative breed of outdoor-enthusiast forester to thrive within the Forest Service. Many (Caufield 1990; Nash 2001; Worster 1993) cite this as the period when public forester Aldo Leopold gained his inspiration. Leopold worked under Pinchot and was one of the first foresters to advocate setting aside wilderness reserves within the public land system. Leopold's (1966) ideas, well represented in *A Sand County Almanac*, captured the ecological and spiritual motivations behind his desire for wilderness preservation. He is considered a founder of wildlife-management science but is more popularly known for his land ethic. Wary of the forestry profession's preoccupation with maximum yields and the capitalization of forests, Leopold insisted that people shift their position from that of being land dominators to that of being members of an overall biological community. His critical 1939 essay, "A Biotic View of Land," published in the *Journal of Forestry,* emphasized "forests as communities of living, interdependent organisms" (Hirt 1994, 42). Leopold viewed human beings as part of this interdependent structure; to destroy any one part was thus to threaten the whole – humanity included. What was imperative, insisted Leopold, was a new "land ethic" that extended the definition of community "to include the soils, waters, plants and animals, or collectively, the land" (Nash 2001, 197). More important, for ecologists, was Leopold's claim that "a thing is right when it tends to preserve the integrity, stability, and beauty of the biotic community. It is wrong when it tends otherwise" (197). On leaving the Forest Service in 1935, Leopold founded the Wilderness Society.

Postwar Years: "Getting Out the Cut"

Initially, Leopold's ideas fell on deaf ears, nor was the growing public interest in recreational uses of the forest taken particularly seriously. The Second World War and the housing boom that followed it generated a tremendous demand for timber. When faced with conflict between preservation and production, the Forest Service invariably sided with the latter. Further, the agency tended to reconcile the need for tradeoffs between opposing public demands on the forest by adopting a blind spirit of optimism about the future timber output promised by "intensive" forest management (Hirt 1994). As this demand coincided with the previously anticipated timber famine on private lands, cutting on public lands increased. Although gaining access to remote areas was difficult and, at first, resulted in a limited harvest, technology soon intervened. The arrival of chainsaws, diesel engines, trucks, and roads in the woods meant an escalating capacity for the rapid cutting, hauling, and milling of timber. More recently, the introduction of processing machines (which can quickly harvest and strip smaller logs) and the application of computer technology to milling have increased technological efficiency. This "efficiency" has accounted for a vast reduction in available timber employment (Robbins 1988).

Except for some market fluctuation, the demand for timber in the 1950s and 1960s continued to climb, although, for individual loggers in individual towns, intensive localized cutting ensured the continuation of boom-bust cycles. In 1946, less than one billion board feet of timber was cut from the national forests in Washington and Oregon. By 1968, the annual cut had skyrocketed to 5.1 billion board feet, reaching 5.6 billion board feet in 1987 (Dietrich 1992, 74). The Forest Service estimates that, out of an original twenty-five million acres (10.1 million hectares) of old growth, only 5.4 million acres (2.2 million hectares) remain in the public trust.[18] In the late 1980s, a Wilderness Society inventory found this to be a gross overestimate. Its findings led to a figure of 1.5 million acres (607,500 hectares), about one-fourth the agency's estimate (Caufield 1990, 65).

Remarkably, as of the year 2001, 57 percent (or sixty-one million acres [24.7 million hectares]) of Oregon's land is owned by the federal government and is controlled by either the US Department of Agriculture's Forest Service or by the Department of the Interior's Bureau of Land Management (Secretary of State 2001, 190-95). Approximately 27.5 million (11.1 million hectares) of these acres are forested, though not all of them are capable of producing timber for commercial harvest. The Forest Service and the Bureau of Land Management are responsible for all timber sales on federal public lands. The vast majority of the commercially viable (timber) lands are administered by the Forest Service via Oregon's thirteen different national forests. The Bureau of Land Management, whose territory is primarily the drier eastern rangelands, does, however, administer the very timber-productive Oregon and California lands. These lands were returned to the government when the Oregon and California Railroad defaulted on promised rail routes. Generally speaking, the Bureau of Land Management has a worse reputation for forest management than does the Forest Service.[19] Nor should the impact of congressional legislation on timber extraction be underestimated. House and Senate representatives have often staked re-election on the generation of federal bills designed to promote timber jobs and tax revenues by selling timber from public lands.

Timber's Workforce

Colloquially, the use of the imprecise, though heuristically useful, distinction between "big" and "small" timber is common. Big industry includes some of the most powerful and affluent multinational corporations in the global arena. They are the corporate descendants of late-nineteenth-century, early 1920s timber barons, who went on to found, among other giants, Weyerhaeuser, Boise-Cascade,[20] ITT-Rayonier, Georgia Pacific, Willamette Industries, and Louisiana Pacific. Their corporate legacy and their accompanying political power is immense; however, as an employing industry, their legacy is waning. In 1950, 62 percent of Oregon's manufacturing workforce was employed by the forest products industry, producing lumber, paper, and wood products. By 1998, only 21 percent of the state's manufacturing workers were employed by the industry. The Office of the Secretary of State attributes roughly one-half of this decline to loss of employment in forest products and the other half to the increase in employment provided by other industries. By 1995, high-tech manufacturing surpassed the forest industry with regard to total number of employees and total overall payroll. As of 1998, 28 percent of all manufacturing workers were high-tech employees, and 75 percent of them were located in the Portland area.[21]

The non-corporate, small business, and worker segments of the timber industry have little economic and political clout compared to their corporate counterparts; however, the two are related in important ways. Both mill owners and corporate timber hire their own labourers and contract with independent loggers, cutters, and logging truck drivers to move the logs from the forest to the mills. Small mill owners procure their logs by bidding on Forest Service and Bureau of Land Management timber sales when any land they own privately becomes exhausted or starts providing a less profitable harvest. Bidding is an expensive and timely process; several years can pass before an accepted bid (deposit included) is legally cleared for cutting. Periodically the expense is more than a small mill can carry, thus small mill owners often subcontract with big timber by purchasing standing timber that has already cleared all financial and legal hurdles. In this sense, small players and workers are often caught in a paradoxical bind as they both compete with and depend upon big timber for employment.

Organized labour, which represents many of the millworkers but not, by definition, independent gyppo loggers and small contractors, supported President Clinton's election in 1992 and again in 1996. Meanwhile, the nationally powerful AFL-CIO supported, both on record and at rallies, the plight of Oregon's timber-affiliated unions: the International Woodworkers of America (IWA) and the Western Council of Industrial Workers (WCIW). But, by the mid-1990s, the membership of both the IWA and the WCIW had fallen dramatically in Oregon. The IWA, for instance, lost nearly two-thirds of its members between 1979 and 1994 (Robertson 1994). Consequently, the unions were not a formidable force during the years of the spotted owl crisis.

Most aptly attribute the unions' diminished presence to structural changes in the industry and to the 1980s market (not to mention the impact of Reagan's overtly anti-union presidency). But other influences are also relevant. For example, as early as the 1940s and 1950s, timber industry officials were fully aware of, and sought to undermine, union-consolidated power. The effect of that effort remains. In the aftermath of the brute force used to suppress the increasingly powerful (and activist) Wobblies in the early decades of the twentieth century, industry officials switched tactics, engaging, instead, in a public relations blitz that fused with, and greatly encouraged, the emergence of timber-related festivals and rituals in the rural west (Walls 1996). Popular to this day are the Paul Bunyanesque sports festivals, wherein individual loggers compete in tree climbing, cross-cut sawing, log rolling, and so on. The events were (and are) very un-labour in that "no unified mass of marching workers" protesting corporate power is evident; rather, "the effect of the sports show was to ennoble the logger symbolically but simultaneously subvert loggers' solidarity as union members and to inculcate the values of competition, rugged individualism, and hard work" (125).[22]

The decline in union presence, alongside the history of public ritual in the rural Pacific Northwest, is currently evident in the very active support grassroots timber activists received from the Gyppo Loggers Association (GLA). The GLA is the trade organization responsible for representing gyppo loggers and their small mill- and woodlot-owning counterparts: it is the communication link and representative body for independent contract loggers. Often at odds with the unions, gyppo loggers are "a different breed." They are not, according to a former forest planner, "big timber." "They are not Weyerhaeuser. They are not Boise-Cascade. These are family-owned small businesses – independent S.O.B.s, every one of them ... and really nice. I mean if you were one of them, they'll do anything for you" (Porter 1996, 114). The fierce independence of many gyppo loggers and small contractors was, as it turns out, a natural fit with the populist logger-identity-based movement of the 1990s.

Early Biodiversity Management and the Wilderness Wars

In 1960, the first of a set of legislative acts meant to represent both environmental and industry interests was introduced. Under pressure from environmentalists and recreationists, the Forest Service came out with a new mission statement: the Multiple Use Sustained Yield Act, 1960.[23] Multiple uses incorporated outdoor recreation, range, timber, watershed, wildlife, and fish interests. The act is notable for its recognition of non-commodity uses of federal forests and for recognizing wilderness, however obliquely, as an idea consistent with a multiple-use framework (Hays 1998, 133). However, the act's lack of clarity left the Forest Service with considerable discretion regarding timber sale and resource decisions (Booth 1994, 144-45). The timing coincided with two events that are singularly important to modern environmentalism: (1) the posthumous publication of Aldo Leopold's 1959 article on biodiversity[24] and (2) the publication of marine biologist Rachel Carson's (1962) *Silent Spring* – a work that has been called the basic book of North America's environmental revolution. *Silent Spring*'s stirring argument exposed the actual and potential consequences of using the insecticide dichlorodiphenyltrichloroethane (DDT). Carson's work continues to be cited in the inspirational biographies of environmentalists, and it spurred dozens of environmental groups into action. Indeed, some of these groups – in the shadow of Muir's legacy – were central to the passing of the Wilderness Act, 1964.

The Wilderness Act, the first substantive departure from traditional forest policy, permitted Forest Service lands to be withdrawn by Congress from the timber pool upon their being reviewed under the Roadless Area Review (RARE) plan (Hays 1998, 133). Wilderness advocacy groups proposed numerous areas they deemed worthy of the wilderness designation. The Forest Service was, in turn, given ten years to designate wilderness areas. By 1972, less than 22 percent of the fifty-six million acres (22.7 million hectares) recommended for wilderness were considered even worthy of further study, never mind protection. It was not until 1984 that the RARE II studies resulted in actual wilderness set-asides in Oregon, Washington, and California. (Almost forty years had passed since Leopold's request for wilderness reserves within the public land system.) The Forest Service defended the delay, saying it took time to identify pure, unadulterated wilderness. Wilderness advocates dismissed the postponements as self-indulgent posturing, accusing the agency of hiding behind "purism" in order to limit the total land eligible for wilderness (Booth 1994, 151-66; Durbin 1996, 31-33, 56-63).

Enter the Northern Spotted Owl

The northern spotted owl first emerged as a dominant concern in the forest dispute in the late 1980s. However, this would not have been possible had Congress not passed three central pieces of legislation in the 1970s: the

National Environmental Policy Act, the Endangered Species Act, and the National Forest Management Act. The National Environmental Policy Act (signed by President Nixon in 1970) requires environmental impact statements for environmentally disruptive "federal actions" (timber sales included). It also requires that the public be included in the decision-making process (Durbin 1996, 32; Robbins 1985, 259). The National Forest Management Act, 1976, emphasizes a balanced consideration of (1) the forests' multiple uses and (2) the need to protect the diversity of plant and animal species. The Forest Service further specified the act's implications by interpreting it to mean that biological diversity was paramount. This was expressed as the "need to maintain viable populations for each tree species" (Hays 1998, 142). Many read this as a sign that forecast the end of Douglas fir monocropping. During the same period, the crucial Endangered Species Act, 1973, provided the potential to protect all species and their habitats despite socio-economic consequences. The ESA thus became the main legal tool for environmental groups and ushered in today's era of environmentalism, which is distinguished by relying more upon scientific (especially biological) argument than upon the need for wilderness preservation per se (Durbin 1996; Hays 1998).

The northern spotted owl was on the first list of threatened species that accompanied the Endangered Species Act. Fiercely contested 300-acre (122-hectare) owl reserves were initiated in 1977 despite resistance on the part of many Forest Service employees and loggers. Discontent escalated dramatically in the 1980s, when radio telemetry studies of owl habitats found that reserves needed to consist of at least 1,000 and as many as 10,000 acres (4,050 hectares) (Dietrich 1992, 76-79; Durbin 1996, 48). Pacific Northwest environmentalists were aware of the implications of the new studies, but they doubted the general public's ability to support massive timber restrictions on behalf of an owl. They were forced to reconsider, however, when, in 1987, a small Massachusetts group (GreenWorld) filed a petition with the Fish and Wildlife Service to formally list the owl as an endangered species (Brown 1995, 29). Initially (1987), the Fish and Wildlife Service claimed that the owl did not need to be listed – a decision that was later reversed (1989) due, in part, to the growing pressure of an increasingly vocal and financially supported environmental community (Durbin and Eisenbart 1993, 10).[25]

Regardless, once a species is listed, the Endangered Species Act requires the formulation of scientifically defensible recovery plans. Government and academic biologists dismissed as inadequate the first plans for the recovery of the spotted owl. Therefore, an increasingly impatient environmental community initiated a flurry of lawsuits that culminated in US district judge William Dwyer's 1990 and 1991 decisions to block all Forest Service timber sales.[26] He ruled that no sale could proceed until an adequate plan for the protection of the spotted owl was approved by the court (Dietrich 1992; Durbin and Eisenbart 1993). Since 1989, 187 Pacific Northwest mills – 40 percent of the then existing mills – have been closed. Industry analysts estimate that 22,000 timber jobs have been lost; meanwhile, the Clinton administration provided 1.2 billion dollars to retrain workers and to assist depressed rural communities (Porter 1999, 4). Despite industry records, the questions concerning the number of jobs lost and the extent of the impact of decreasing timber employment in the region remain open. Some argue that the expanding 1990s economy was sufficient to offset job losses; others argue that there are many economies in Oregon and that many rural communities were especially devastated.[27]

The Clinton Administration

President Clinton inherited the responsibility for coming up with an adequate owl protection plan after the first Bush administration resisted the implications of a conservative plan involving extensive logging restrictions. The 1993 Forest Summit produced the Clinton administration's first effort to draft a comprehensive, "scientifically defensible" forest plan that would protect the owl along with other potentially threatened species. Judge Dwyer's injunctions and the final Northwest Forest Plan's promise of a comprehensive restructuring of forest policy shifted the focus from the protection of individual species to ecosystem management – a shift that, ostensibly, would take into account logging's impact on all species. Pointedly, the plan proposed reducing annual harvest levels by 80 percent – from the five-plus billion board feet reached in the 1980s to approximately 1.2 billion board feet – the largest scientifically defensible harvest on federal Pacific Northwest forests. The backroom lobbying and legal wrangling that engulfed the attempt to achieve an "acceptable plan" were described by all sides as bitter, draining, and discouraging.[28]

Many insist that the Clinton plan is working, but its record is, at best, mixed. It was stalled dramatically by the conservative shift ushered in by the 1994 federal election, which brought with it numerous Republican victories based on overtly anti-environmentalist platforms. That same year record wildfires swept the west, destroying four million acres (1.6 million hectares). Under pressure from a Republican majority in Congress, "the Administration officials saw salvage (logging) as a way around their earlier tough talk of 'science-based' forestry policy, a means, under the auspices of an emergency, to find a few billion feet of hassle-free timber for noisy constituents out West" (Roberts 1997, 49). Charles Taylor (R-NC) produced a "salvage rider," more formally known as the Federal Lands and Forest Health Protection Act, 1995 (Roberts 1997, 257). Under the provisions of this act, salvage timber sales, easily defensible on the basis that they eliminated diseased trees, were exempt from appeal under extant environmental laws.

But the act's definition of salvage logging included "dead and dying" trees as well as associated "green trees" (Roberts 1997, 49). Blocking the salvage act in Congress was not possible because it was tacked on to universally sympathetic legislation; namely, reparations legislation for the Oklahoma City bombing (see the Emergency Supplemental and Rescissions Act, 1995). Even Peter DeFazio (D-OR), the notoriously even-handed representative from Oregon, criticized the act for defining salvage so broadly "that it opened the door to wholesale logging in the region's remaining old-growth forests and roadless areas" (Durbin 1996, 257). A new series of bitter direct-action protests ensued at salvage sale sites, while others campaigned locally and nationally to prohibit "logging without laws." Finally, as 1996 drew to a close, Clinton rescinded the rider, a gesture made possible by the moderating forces that re-elected President Clinton and Vice-President Gore that same year.

President Clinton's finalized 1994 Northwest Forest Plan nonetheless reigns as the region's key public lands forestry policy document – a document in which the language of logging has shifted from "allowable cut" to "ecosystem management." In the eyes of both timber advocates and environmentalists, the plan's long-term implications remain uncertain. Environmentalists are cynical about past and future salvage logging, while timber-dependent workers are still suffering in the wake of legal bans on timber sales. Further, the plan has not resulted in an end, as once hoped, to lawsuits, appeals, or court injunctions. In the summer of 1998 Judge Dwyer blocked 400 million board feet of proposed timber sales because the Forest Service failed to survey for rare plant and wildlife species, and in August 1999 a judicial ruling halted nine federal timber sales in the Pacific Northwest for the same reason (Murphy 1999, D2). Most recently, Oregon's Office of the Secretary of State (2001) commented that the decline in harvests on federal lands within the state's borders had plummeted from highs of several billion board feet to a much reduced 383 million board feet in 1999. Yet, this could well change under President Bush. The new under-secretary of the Department of Agriculture, who is responsible for overseeing the Forest Service, has pledged to deliver the full volume of timber allowable under the Northwest Forest Plan (Harwood 2002).

There is an undeniable ebb and flow to this historical sketch. The social, biological, and political events that unfolded during the colonization of the west involved a series of gains and losses. Much has irrevocably changed, some things have not. Land use in the Pacific Northwest did not always imply ecological devastation; there is another history (Aboriginal and non-Aboriginal) that we have begun to learn from all over again. Selective logging, once conducted purely due to technological restrictions, may now hold considerable promise for those promoting minimal, low-impact logging. A look at the legislative history of the region reveals both a recurring tolerance for illegality and a strict application of some of the most biologically protective laws in the world. Theodore Roosevelt is still "the conservation president," but Roosevelt-style conservation is closer to what today's loggers would support than it is to what today's environmentalists would support. Popular though it is right now, the concept of sustainability is not new to the timber industry, to public lands policy, or to environmental disputes over how to manage the forests. The loggers' belief in an activist populism and a continuing forest that will replenish itself indefinitely can be traced directly to the Roosevelt-Pinchot legacy. Meanwhile, the roots of modern-day environmentalism can be traced most prominently to John Muir's spirituality, Leopold's land ethic, and the ecosystemic sensibilities articulated by George Perkins Marsh and Rachel Carson.

3
Disturbances in the Field and the Defining of Social Movements

The Ancient Forest Movement and the Forest Community Movement were the primary local grassroots interests attendant to Oregon's old-growth dispute at the time of my research. That they have ties to national and international bodies goes without saying. The striking feature of both these groups, however, is that they do not conform to traditional socio-political divisions based on class, gender, or political party. Such structural and political affiliations infuse both movements, but their overarching feature is their collective self-naming and identification as loggers or environmentalists. In other words, beyond any given set of structural characteristics, it is the self-reflective pursuit of recognition as a collective identity that defines these movements (Melucci 1994). Designated by social scientists as "new social movements," their identity-seeking acts are often organized around the threat that other social groups, institutions, and actions pose to their self-definition and "right" to exist as loggers, environmentalists, pro-choice (or anti-choice) activists, and so on. The boundary between groups is maintained by the constancy of we-they distinctions, and, as discussed in Chapter 1, by the competitive (and dialogic) process of sustaining and crafting meaningful social spaces wherein identities can be experienced and defined (Johnston, Laraña, and Gusfield 1994, 10-20).[1]

The Forest Community Movement
It is common to think of activist loggers and ranchers as part of the Wise Use Movement (or, worse, the militia movement). Wise Use came of age in the 1980s alongside the Sagebrush Rebellion of the 1970s and 1980s (Cawley 1993). Closely associated with Ronald Reagan's former interior secretary James Watt, Wise Use is best known for its desire to privatize public lands so as to render them open to widespread resource extraction. Certainly, there are and were corporate interests who might benefit enormously from privatization; however, despite the rhetoric, loggers, ranchers, and environmentalists have all historically sought federal help to create and maintain a west

in their respective images. "For over a century," notes Hess (1995, 163), "politicians, policy makers, and functionaries have choreographed almost every facet of life" in the west. Privatization is ultimately too risky for either group as neither can fully control who would then end up with "the land" (167). The more precise question is, which west is the federal government going to support? For periods, under the Clinton administration, it appeared as though the environmentalist west might, for a change, prevail. However, under President George W. Bush, an escalating threat to the environment seems certain as conflicts surface in the western United States over the drilling of oil in Alaska's wildlife reserves and federal promises to open public lands to more extensive resource extraction.

Presidential shifts aside, it is erroneous to dismiss loggers (and ranchers) as evil or morally objectionable on the basis of Wise Use affiliations. Many loggers and ranchers are not "wise use": they are simply decent rural people who labour under federal laws that leave them with no other incentive or option but to maximize the per head of cattle or board feet of timber to be extracted (Cawley 1993; Hess 1995). Moreover, institutional policies indifferent to rural Oregon are all too familiar to loggers who have suffered for years under the boom-bust cycles instigated by both federal and corporate entities. For this and other reasons, the central concern of those studied here – members of the forest community movement – is the long-term sustainability of communities built on family-wage employment, a spirited logging ethos, and the forests upon which this depends.

The legacy of Gifford Pinchot-style populism and the principle of the greatest good for the greatest number is endemic to their perspective. Their rallying cries of "people count too" and the need to protect a distinct way of life are oriented towards the principal goals of community well-being (Carroll 1995). Overlap undoubtedly exists between Wise Use and the forest community movement, but the central moral force behind the latter is the right for small logging entrepreneurs (and I do mean small) to earn a living from the forests and to place limits on the dominant economic and political interests that have historically destabilized forest communities.[2] The single best expression of this movement and the source of all but a very few of my logging community interviews and fieldwork contacts is the above introduced Oregon Forest Community Coalition (OFCC), many members of whom are also members of the body that represents independent loggers – the Gyppo Loggers Association (GLA).

The OFCC encompasses approximately sixty Oregon community groups representing timber workers, small businesses, and members of resource-dependent communities throughout the state. The coalition and its constituent groups united in the summer of 1989 after Judge William Dwyer issued the first injunctions on national forest timber sales in spotted owl habitat. They affected those most dependent upon public lands: small businesses and independent logging contractors. Larger companies with extensive private holdings posted record profits because the market value of accessible (i.e., private) timber escalated after the injunctions. This expanded the extant wedge between "small" and "big" timber, and it secured the need for workers and smaller companies to fight their own battles.

The OFCC's headquarters are tucked away in a small decaying storefront about half a mile from Oregon's state capitol buildings. Though industry's resources are seemingly inexhaustible, the same cannot be said of the OFCC's. In 1992, its state-wide membership was estimated at 77,000.[3] Fund raising and donations had produced a $145,000 budget to cover operating costs and salaries for staff members (Durbin 1996).[4] The coalition nonetheless struggled to defend its grassroots status (a subject addressed explicitly in Chapter 4) and, hence, resented media depictions of its members as industry "officials" or "spokespersons." One woman's letter to the editor of Oregon's most widely read newspaper captured this struggle:

> Charlie Janz, a logger who employs 12 and 25 people (two logging crews), is hardly a "timber industry official," as you called him ... The grassroots movement is anchored by people like Janz, by X-L Timber office manager and mother of five Debbie Miley, by single mother and Hanel Lumber employee Rita Kaley, by ranchers and truck drivers and by the 50,000 others who have a personal stake in the use of this state's resources.
>
> If big "I" industry was the answer to our prayers, we'd have left the 65-hour weeks to it ... For as long as people like us have known, hard work was enough. Providing for our families and making our contributions to our schools and communities were enough. But our communities are in danger, and industry is not the answer. More fat-cat lobbyists and timber-industry officials are not the answer. (Sparks 1990, 7)

The Ancient Forest Movement

Like loggers, many local environmentalists also experienced moments that seemed more significant than others with regard to their capacity to galvanize their collective identification as purposive activists. A case in point is the 1986 cutting of Millennium Grove, an event that left many of those prominent in the spotted owl controversy deeply shaken and profoundly distrustful of the Forest Service.[5] Not far from Corvallis, Oregon, Millennium Grove, or the Squaw Three Timber Sale, as it was known to the Forest Service, contained some of the oldest and largest trees in the continental United States. The felling of the old-growth giants was referred to as the Easter Massacre (cutting began on an Easter Monday). The protests that engulfed the sale included the first tree-sitting incidents in Oregon and, more fully, secured the role of civil disobedience at contested logging sites (Dietrich 1992, 157).[6]

The term "ancient forest" was coined by a local activist who appreciated "ancient's" two-syllable elegance and the image of historical monuments it evoked. In the aftermath of the court injunctions, "ancient forest" quickly became a household phrase, and the fate of Oregon's forests became a national concern. The relationship between local grassroots interests in Oregon and national environmental non-governmental organizations (NGOs) was neither intimate nor hostile, merely tense. There was a sharing of both information and political support. Audubon sponsored some of the earliest seminars, provided informational material, and mounted a legal defence of the owl. The Sierra Club Legal Defense Fund's opening of a Seattle office in 1987 provided the legal expertise for a sustained pattern of lawsuits. The Wilderness Society pumped the most money into quantifying how little old growth was left (Dietrich 1992, 214).

Still, members of local grassroots groups often expressed discontent with the swollen budgets and conservative tendencies of the national groups. Moreover, the reputations and coffers of environmental non-governmental organizations (ENGOs) had benefited significantly from events in Oregon, while local groups were left to do the movement's dirty work.

Much of that work was accomplished by the Ancient Forest Grassroots Alliance (AFGA), the oldest, most renowned, and most powerful grassroots group in Oregon at the time of my investigations. Beginning with three staff members in 1972, by 1992 it had grown to comprise a paid staff of fifteen with volunteer offices in Eugene, Portland, and Bend, Oregon. More important, it had grown into a state-wide coalition of fifty conservation groups with 6,000 members.[7] AFGA activists, along with higher-profile participants from AFGA's constituent groups, were the source of all my interview and fieldwork contacts.

Disturbances in the Field

As noted on page x of this book, the majority of the research materials discussed herein consist of the transcripts of recorded ethnographic interviews conducted with individuals involved (either directly or as part of a member group) with the OFCC or with AFGA.[8] Whenever possible, however, I also attended meetings, conferences, and "owl" hearings; toured mills; and visited logging sites, cutting sites, and tree farms. I attended demonstrations; wandered about in the rain-soaked crowds outside President Clinton's Forest Summit in Portland; and, whenever possible, attended speaking events. The following vignettes offer a feel for the type of fieldwork in which I was engaged, and they bring to life those loggers and environmentalists who lived and breathed the conflict. I often empathized with both; I rarely agreed entirely with either.

The political climate in Oregon in the early and middle 1990s was evident in the minute particulars of my surroundings – the headlines in the local paper, a bumper sticker here, a lapel button there. It was difficult to move through public space without encountering some evidence of tension. One Saturday I happened across a Forest Service road while driving to the Oregon coast. It was, I learned, a designated auto-route devised by the Forest Service to educate the unenlightened public they imagined might be out for a weekend drive. Along the roadside was a suite of signs, each – in deference to Gifford Pinchot's legacy – depicting the logger as a farmer and the forest as a crop. Douglas firs were said to grow most successfully on "bare" (i.e., clear-cut) land, and conservation was explained in terms of economic efficiency. Upon return to private property at the edge of the Forest Service road, one encounters three accompanying hand-made signs. They are spaced to provide easy reading for passing drivers, and together they read: "If environmentalists could have their way, we [timber advocates, one assumes] would live from day-to-day. The spotted owl and seal would thrive, and you and I could not survive." At home that evening, I walked to the convenience store for milk. A sign in the first block of wood-framed houses that I passed read: "This family is supported by clean air, pure water, and an intact ecosystem." On the second block, the sign on the bumper of an old Ford truck proclaimed: "This family is supported by timber dollars."

Hearing the Owl

On a blistering June day in Roseburg, Oregon, the Scientific Assessment Team advising the federal government on protecting the northern spotted owl met with concerned citizens. Roseburg is situated in one of the Umpqua River's many valleys and is one of a handful of timber-enriched small cities in southwestern Oregon. Much of the land surrounding the town consists of national and private forest that has long been harvested for the coveted Douglas fir. I had made the ninety-minute drive from Eugene to attend the hearings because they were the only ones to be held in a timber-dependent community that spring. Outside the community college building that staged the hearings, one of Roseburg's larger logging companies had parked one of its trucks. Laden with a full load of old-growth timber, the truck had two large banners spread the length of its right side. The first read: "People Count Too"; the second read: "Preservation Kills." A petition was being circulated calling to have loggers listed as an endangered species.

Sitting towards the back of the theatre, I could see before me a large cluster of women with children. The men sat together towards the auditorium's front end. The hearings consisted of a sequence of twelve speakers, all but one of whom criticized the early owl-protection plan for the immense social and fiscal impact that its widespread logging restrictions would introduce.[9] Contempt for the plan and its defenders was palpable, a discontent voiced most frequently as the need to "see the human side of the picture."

The crowd leapt to its feet when a county commissioner proclaimed, "We are real people, not statistics," following a municipal politician's statement that adhering to the plan would mean the loss of thousands of jobs. The second most emphatic response to any speaker followed the words of a local industry forester, who commented that a "better name for the plan would be the Pacific Northwest Family and Community Genocide Plan." This brought the audience to its feet a second time. Response to the woman speaking on behalf of the Audubon Society came in the form of impatient and contemptuous mumbling throughout her presentation. When she criticized the plan for failing to guarantee the owl's survival, the woman sitting next to me muttered a loud and curt, "So what?"

Hauling Logs

Upon my arrival in Oregon, the world of loggers and others advocating continued timber harvesting on public lands was relatively unknown to me. I could claim to have planted trees in remote clear-cuts during my early twenties in British Columbia, but, otherwise, my life had followed an urban, middle-class course. Thus it was the appearance of a thoughtful letter to the editor in the local Eugene newspaper that provided me with my start. The letter concerned the stereotyping of loggers and was written by Jim Stratton, himself a veteran logger. His whereabouts was traced through the paper; within a month he had introduced me to a handful of timber advocates (whom he repeatedly referred to as "decent men"). He also made arrangements for me to visit the crew and site of his current logging job.

A first shock came with the realization that a logger's day begins in the not yet daylight hours. I'd thus arranged to meet with Jim at 4:00 A.M. outside a convenience store in a tiny Oregon town. There we waited for the crew bus, or the "crummy," as it is known, which we followed to a road at the base of the logging site. Jim cursed the absence of morning coffee (he'd forgotten his thermos), while I fretted over the impression I would convey to those I did not know. Did my jeans look as though they had ever seen a day's labour? Did my rain jacket look new, inhospitable to grime? And what was I going to do about my sometimes paralyzing timidity?

The night before meeting the crew I slept poorly and dreamt of walking down a seemingly endless aisle on a large Bluebird bus, being glared at by burly men in chequered shirts. But the crew bus was not the large yellow vehicle I'd imagined; it was simply an extended pick-up inhabited by five quiet men, most of whom were not yet awake. Jim and I joined them and exchanged grunted introductions. Leaving the base road, the truck lurched and pitched its way up five kilometres of marginal logging road and came to rest on a dusty flat.[10] The flat was the stage upon which was set the massive machinery that hauled and lifted logs and/or processed smaller logs.

During the wait for daylight, members of the crew ate bagged breakfasts, smoked, and laced work boots.

Technically, "logging" refers to the drawing of already felled timber uphill to awaiting trucks. At daybreak, the crew's two youngest men (one a former forestry student, the other a certified math teacher who had yet to teach) set off downhill to set chokers.[11] Chokers are the large cables that are secured around the logs, after which a tower (called a Berger) with an enormous engine at its base hoists them off the ground and up the hill.[12] Setting chokers is logging's most dangerous job, and it is usually assigned to younger loggers. A snagged or broken cable can send several tons of timber hurtling downhill, resulting in what one woman referred to as a "choker widow." A third man (Rick) stood atop the hill directing traffic. He was the "chaser" – the person who communicates with the machine operators and choker setters. A series of hand signals and shrill whistles are used to initiate the movement of logs and cables. Jim runs the Berger, the most coveted on-site job, while another man operates a log processor that strips and stacks smaller (40 to 50 centimetres in diameter) logs.

I spent most of the day watching and occasionally talking with Rick, but conversation was generally out of the question due to the deafening roar of the machinery. Most of the crew members had already suffered significant hearing loss, rendering easy conversation impossible. Eventually, Jim offered me a pair of malleable yellow earplugs, an accompaniment to my bright orange hard hat. Were it not for the omnipresent swarm of bees, comfort might have been at hand.

In time I settled onto a perch atop a pile of log debris. The earplugs transformed the noise into a hypnotic hum that numbed my awareness of the surrounding destruction and made it easier to enjoy the distant mountains. The patchwork shades of green and brown on the surrounding hillsides indicated the large proportion of visible land that had been clear-cut in the last fifty years. Because we were on the western edge of the Willamette National Forest, some of the acreage in view included private land where the size and volume of clear-cuts often dwarf that on public lands. Deep scars were evident on recent clear-cuts, a web-like pattern carved by the cables that stabilized the tower portion of the heavy machinery.

When the machines were shut down for a brief period to be oiled and cooled, the hum of similar equipment resonated throughout the valley. Equally audible was the crashing sound of trees felled by cutters in adjacent sites. This incredible ripping sort of noise was followed by a heaving, rattling thud and subsequent echo. The experience left me agitated, spooked. To distract myself from this, I turned my attention, again, to Rick, who was eyeing the smaller processed logs to determine which would become telephone poles, which two-by-fours, and which would be designated for paper mills. Part of his task was to weed out logs with "sucker" knots – knots on the outside of logs that draw water into the wood's core, thus causing rot.

Rick was nearly fifty at the time, an avid outdoorsman who had worked in logging all his life. He believes that there is plenty of timber for everyone. "Non-loggers," he said, "can't see past the clear-cuts, don't enter into them to see that they've been replanted." When I asked him what defined old growth, his response was indirect, sarcastic: "Whenever you try and bid on a sale of old growth it gets tied up in court." To Rick, old growth is whatever environmentalists are willing to fight over in court. Although he applauded wilderness "set-asides," he insisted that environmentalists were "expecting too much."

Rick then told me a story of a logger he knew who had camped near a stand of old growth. An adjacent "wasting, rotting" log fell; the logger was able to get his children out of the way, but he lost his leg doing so. Rick believed that it would take loss of life under like circumstances to "bring public awareness around" to seeing the futility of permitting old growth to stand "rotting, unharvested." Though not especially credible, Rick's story stands as an example of his attempt to balance the value of owls against the value of humans. He appeared to be asking why the possible extinction of the northern spotted owl can generate such public outcry when the possible death of a logger-cum-camper cannot. Many have argued that loggers "choose" their lifestyle while owls do not, but loggers do lose their lives at an alarmingly frequent rate while cutting and hauling timber, a fact that has garnered little public sympathy.

As the day grew hotter, from his side pocket Rick extracted a cylindrical tool used to gauge the air's moisture level. If the humidity falls below 30 percent, then logging crews must shut down in order to avoid the risk of fire. Conversation during a brief lunch break revolved around derogatory jokes made at each other's expense. The choker setters, whose college experience I have already mentioned, were repeatedly referred to as "dummies," while Dan (the man who operated the processor) was ruthlessly teased about his thinning hair. The crew also took turns berating their absent owner/boss, Bill, for his stinginess. The crowning story concerned Bill's third child, who was said to have lived with relatives because Bill was too cheap to add a room to the tiny cabin in which the family lived.

Conversation about Bill turned to conversation about the "glory days" of logging – the era of endlessly available work. In the late 1970s, a relatively fit White male could feasibly stand on any corner in any timber town and find crew work (Dietrich 1992). In Rick's words: "It used to be you could log for a while, quit your job anytime, and have a job an hour later that day if you wanted it." Rick said that for years he would log through the summer, take the winter off (fishing and collecting unemployment), and return to work in the spring.

After lunch I accompanied Rick to pick up fuel for the machinery. He tested my impressions with a comment that was part challenge, part inquiry: "I bet you didn't expect loggers to be nice people?" He was clearly pleased when I told him otherwise (somewhat dishonestly, given my previous night's dream). We spoke of the city in which I lived, a place Rick views as "pretty much full of environmentalists." But Rick followed this comment with a compliment, a covert reference to myself as being unlike the rest of my urban neighbours: "All those people in the city who sit in their offices and decide whether or not we can cut areas. And how is it that they never actually come up here and see ... see that it all grows, it's all fine?"

When we returned to the landing, the spectre of extreme fire hazard had set in. This work day would end early. On the way home in the crummy the objects of the men's jokes were owls and environmentalists: "Better hide the two quarts of motor oil we spilled today from the environmentalists;" "There goes that plain old barn owl making eyes at that spotted owl." I enjoyed the joking and the apparent tolerance of my presence. About two weeks later, Jim told me that the crew had enjoyed my visit and had appreciated the fact that I "listened" and "didn't talk all the time." By way of thanks, I was given an orange-billed cap advertising heavy equipment for logging.

Cutters

As the cutting of old-growth timber is the crux of the forest dispute, in the spring of 1994 I sought and found an opportunity to watch cutters at work. The day began at Sanderson Brothers, a small old-growth mill, one of many rural operations dependent upon the timber-rich Willamette National Forest. There I met with Doug, who seemed oddly surprised that I had shown up. We had met several times previously, when I had interviewed him at length. Doug is a third-generation logger, a tall man in his late forties, bearded, greying, and balding. Having worked his way up the millworker-logger hierarchy, he was employed by the mill to supervise logging crews and to "cruise" for timber. Cruising involves searching for, and estimating the potential value of, individual timber sales. As on every other day we had met, he was wearing the logger's pinstriped work shirt, red suspenders, and jeans cut short to resist snagging on forest undergrowth. All that revealed his role as foreman and timber cruiser were the pens and note pad in his breast pocket.

Our first stop was a local convenience store and deli – coffee and snacks for Doug, a bag lunch for me. He insisted on paying, as I was his "guest." "Chivalry," he added, "is not dead at Sanderson Brothers," a statement that served as a clear reminder of my gendered position. We headed west on the state highway and eventually turned off on one of the paved Forest Service roads in the Willamette National Forest. These paved roads attest to the volume of logging traffic and the commercial wealth of the forests through which they cut. The Forest Service has constructed 343,000 miles (552,000 kilometres) of paved, gravel, and dirt roads, a distance that is seven times the length of the US national interstate highway system (Dietrich 1992, 63). At lower elevations the ground was covered with young, spindly alder, which appeared, according to Doug, after a logging operation twenty-five years earlier. Doug referred to this messy, light-barked forest as an example of "mother nature doing a good job."[13] A black belt in karate, Doug also spoke of his beloved martial arts and his interest in Eastern philosophy. In his spare time, Doug teaches self-defence classes for women. (I noted that he said "women," not "girls" or "ladies.") Doug, in his boss's words, is "a square peg in a round hole."

When we arrived at the top of the highest plateau in the area, Doug remarked that "as far back as the 1950s there was talk of Oregon running out of wood." Envisioning a continuous forest, he commented on the growth occurring in every green patch and replanted clear-cut. On two occasions Doug averted my attempts to talk not about growth but about old growth; he was much more confident when talking about timber supply, about the ebb and flow of board feet.

I was taken aback by the steepness of the grade on several of the clear-cuts and said as much to Doug. He responded, "Those hills are steeper than the back side of God's head." Doug used God and Nature interchangeably, saying on several occasions that "Nature" and/or "God" does a better job than anyone; "Nature has a healing power beyond anything we can imagine ourselves." Nature, in Doug's conception, is robust and can tolerate most human impact. For my own part, I very much wanted to see the smooth cycles of growth and continuity that Doug saw, I wanted to see his cyclical, returning forest. But I did not. Instead my eyes were repeatedly drawn to the intermittent acreage of scarred and denuded hillsides.

As we gained elevation on the now gravel road, Doug spotted something on his side of the truck. He peered over the road's edge, stating that there could be a "herd of elk" in the clear-cut below; I could not spot the herd myself. Elk follow the grazing opportunities made available by the cleared land's new growth, feasting on chest-high saplings and brush. I wondered whether Doug was putting on a display for me by mentioning elk. Loggers tend to talk of big game, what environmental ethicists call "charismatic megafauna," while environmentalists tend to talk about lichens and fungi, forest minutiae.

The first cutting site we visited that day was just ahead. I didn't see the cluster of three giant trees slated for cutting until I was out of the truck. "Fifteen old-growth trees come down on a good day," announced Doug as I tried to keep my low-grade anxiety at bay. The cutters looked tiny from where Doug and I stood. The forested area in front of me that had just been cut but not yet burned or logged was a comparatively attractive sight. Everything was low to the ground but still very green and dense. I couldn't help thinking that if this was what clear-cuts finally looked like, the more typical jarring visual impact of aerial clear-cuts with their dried and burned land would be less beneficial to the environmentalist cause.

The cutters followed two downed trees and severed them with chainsaws into the lengths specified by the contracting mill. This process is referred to as "bucking." Doug remarked that many downed trees roll instead of landing in the intended spot. These have to be chased by the cutters to ensure that they are bucked and accessible to choker setters. Before leaving, Doug and I watched one giant come down and crash with great force. The ground beneath our feet trembled when the tree bounced before coming to a stop. For the first time that day, I didn't see a gouged and disrupted forest but, rather, a dissected garden. Given this fragmented landscape, the notion of interconnected ecological systems seemed quite removed.

Cutting had just begun at the second site. We parked beside two other trucks, thus creating a roadblock to keep stray travellers from the path of falling trees. One of the trucks, a Chevrolet Blazer, had a bumper sticker that read, "Do you work for a living or are you an environmentalist?" A young man was working by himself, clearing brush and small hardwoods. When Doug told him I'd come to observe cutting, he told me that he was "not the one to watch" as he was "new at this game." His tin hard hat (like the one I wore) was garnished with a fluorescent hatband that read: "killer trees." This says about as much as any two words can about cutting: cutting is always about felling a tree so that it doesn't land on the cutter. Flying branches and debris can easily maim or kill (as can, of course, the falling tree itself).

Doug and I scrambled down the hillside to where a pair of cutters was working on a very steep pitch. The older man, Sonny, was friendly and talkative, the younger man was stoic, quietly smoking off to the side. Doug introduced me as someone "trying to put a balanced angle" on the dispute, a characterization I don't recall offering myself. The older man had "nothing against putting aside some of the more magnificent stands," but he insisted that the controversy "needed some middle ground, some reason." He struck me as eager to be seen as sensible, moderate – a touching and familiar stance.

When the pair returned to work, Doug and I remained perched above them. Before a tree comes down, the area around its base is cleared of anything that might obstruct it and change the course of its fall. A "face" (a large curved wedge that looks like a smile) is cut from the side of the tree in order to aim the fall. The cut must be precise; a minor deviation in angle can mean a fall in a potentially fatal direction. After the face is cut, the fallers call out three times (the classic "tiiimmberrr," or anything else, for that matter) to warn those in the vicinity. Cutting then begins on the side of the trunk opposite the face. When the tree begins to fall, the cutters hustle out of the way. On steep terrain cutters hope to drop the tree horizontally into the hillside to minimize its chance of rolling. The tree is said to have "bucked" when it inadvertently hits something in its path, causing breakage in one or more places. Steep ground and trees rolling downhill are what make the difference between a five- and a fifteen-tree day per pairs of cutters.

Before the fall of the first tree at this site, Doug asked me if I knew my "escape route." What would I do if the tree didn't fall in the intended direction? I agreed to move towards the protection of the humped indentation behind me. If, on the other hand, I were to spot a falling branch, I was "not to turn and run but to watch the branch fall and jump out of the way at the last minute." (This plan seemed counter-intuitive, like the admonition to drivers to turn into and not away from a skid.) As we waited, conversation turned to the lack of monotony in this work. Doug grew serious: "The moment a cutter ceases to think could easily be his last."

At one point Doug announced that "this is as good as it gets." I assumed he was reminiscing about his days as a cutter and logger. His meaning was clarified when he followed with, "You can't get any closer to the universe, to nature, than this." I was a bit taken aback and could not help recalling the anti-military bumper sticker: "Go to exciting places, meet exciting people, and kill them." Again, I kept my thoughts to myself.

When Doug turned to leave this site, he moved quickly upward through the dense brush. I assumed my role was to follow as best as I could and did so. It was not my legs that bothered me, but my wind. At the top of the ridge Doug mentioned the need to get out of the way as soon as possible. "The cutters need to see that we're out of the way, out of their range of responsibility." Doug was impressed by my ability to keep up with him, called me a "trooper," and compared me favourably with a woman from the Forest Service "who just couldn't keep up." The compliment pleased me, as I quite enjoyed Doug's spirit and appreciated his knowledge of this world.

Propped up against the side of the truck, Doug spoke of the numerous logging accidents he'd seen: "bodies crushed and dismembered." Among loggers, I have learned, there is always talk of danger. They all know someone who died "working the woods." Most fallers are injured at some point in their career: they just hope the injury won't be fatal. I asked Doug if he had ever been injured. "No, but only because of luck." I asked him if he ever considered quitting, having seen so many accidents. He said that when his best friend was killed by an out-of-control log, he very nearly did quit. He had the same reaction when "a kid just out of high school was crushed to death by a log."

Over the course of the day, discomfort comes and goes. I loved being outside, but I felt as though I could not possibly absorb the conflict between the two worlds with which I was in contact. On our way to the last cutting site, Doug made reference to protesters hooked to fences on logging roads. He told the story of a young man who had used a bike lock to attach himself to the road's fence. Doug swung the gate open while the protester choked and his girlfriend screamed. Sardonically, Doug added that he had to slam the gate shut. At moments like this, I would become silent or mumble some neutral comment.

The last cutting site of the day was unlike any of the others – perfectly flat, easy-to-negotiate land. We approached Sonny first; he had finished with the earlier cut and relocated here. He was small, slight, and seemed too old for this work. He joked about the forest dispute giving him all his grey hair and said that he was really only twenty-nine. His young stoic cutting partner did not join in the introductions, preferring to stay at his job, ear protection intact. The jokes about the ease of cutting at this site continued for some time, particularly in the form of melodramatic complaints about the logs these cutters had to chase downhill. (There was, of course, no downhill.) In jest, Sonny flashed his wallet and suggested paying Doug to keep the other cutters from knowing where they worked. Light-duty cutting, which is really anathema, is a rare treat for cutters, a kind of guilty pleasure.

It was at this last site that the felling of timber bothered me the most. The first tree fell exactly as planned, but I was certain I heard a piercing, human-like shriek as it went down. Momentarily, I wondered whether the faller, the silent one, had screamed for my benefit, embarrassed though I was by the thought. Nonetheless, the sound echoed in my head for the remainder of the day. I was a carnivore who hated hunting for food, a consumer of wood products unable to reconcile myself to their source.

Before heading home we stopped at a logging operation. Two enormous logs were aerially suspended by a cable and attached to a "sky car" (a bit like a small gondola) that carried the logs up the hill via cables. It looked like a surreal game of pick-up sticks until I heard the sky car's two logs careen into each other. A hollow booming sound echoed throughout the valley, and the already mighty and thick cables heaved as the logs swung. No one moved. Doug asked me if I would ride the sky car, explaining that crews used to ride the cars uphill (by simply hanging on) at the end of the day. This practice came to an end when a crew in Washington State was killed when a cable snagged and then released, reverberating brutally and tossing them to their deaths.

Doug's company truck screeched all the way into town. Worn break pads were grinding on metal, testimony to hard financial times at the mill. One of Doug's comments during the drive home stayed with me, typical as it was of many loggers' appeals for sympathy: "Loggers," he said, "are like eggs; tough on the outside, soft on the inside."

A Roadside Protest

Later the same year, I joined a gathering organized by Oregon Eco-Defenders. A camp had been set up in a roadless area deep in one of Oregon's national forests. The area, identified as owl habitat, was slated for helicopter logging in the immediate future. The standing timber had been sold before the legally enforced suspension of timber sales and was, therefore, not part of more recently proposed owl habitat.[14] When making the arrangements for this excursion, I was careful to establish myself as an interested researcher rather than as a committed group member. Nonetheless, organizers were welcoming and assured me that I need not "belong" in order to attend.

I set out on a cool, clear morning and picked up an attending activist (Michael) to whom I'd promised a ride. He worked for a group of organic farmers outside Eugene, had never been to an on-site protest, and also professed to be interested in meeting other people. About noon we pulled up alongside the cluster of vehicles haphazardly parked on the logging road closest to the sale. Camp had been pitched on the only flat on an otherwise steep, forested slope. The cars and trucks were strewn with banners that read "No Compromise with Mother Earth," and "No Deals/No Compromise." About twenty people were already at the site when I arrived, most of them from urban areas in western Oregon and a few from a college town in northern California. About two-thirds of the group were male. All but two were "twenty-something." Later, this group was joined by about ten local activists. The average age of this second group appeared to be about forty-five, and most of its members had made the conscious decision to move to rural Oregon during the back-to-the-land movement of decades prior. Overall, both groups reminded me of the White middle-class rock-climbers, mountaineers, and hikers I had worked with in retail sales while an undergraduate. The mood was distinctly social and recreational; something of a reunion was under way for those who had attended a similar gathering the year before.

Saturday's loosely organized agenda involved a short lecture on the details of the timber sale, as well as hiking expeditions that were meant to familiarize participants with the threatened area. Tree-climbing lessons were available throughout the day. Sunday was to involve a non-violent, direct-action workshop and a strategy-planning meeting. That first afternoon I set up my tent and made arrangements to go on a hike with Alan (a college student) and Tess (a wildlife biologist); I had already identified the latter as the group's unofficial leader. Tess organized most of the on-site activities and lectured interested persons on the sale's boundaries and ge-netic distinctiveness. She was a smart, no-nonsense woman who earned my respect immediately. The three of us worked our way up the hillside that had the largest volume of trees dedicated for cutting. At the top of the hill I got my first 360-degree sense of the area. It had been a hot, dry summer and fall, consequently, the air was thin and sweet with the smell of parched pine. There were no discernible logging roads from this point – a rare visual feast. The quiet was otherworldly. Deep in the forest the trees creaked and swayed (a slow metronome-like movement), while the winds rustled every-thing. The crows, as usual, were noisily annoyed with our presence. Water was scarce and several creeks were dry, a pattern that does not change until late November.

Back down the hill, Tess pointed out a few rare species of cedar. This part of Oregon was not affected by the last glaciation, hence its atypical concen-tration of species. It is also, Tess noted, a biological corridor – part of an uninterrupted stretch of old-growth forest through which the wide-ranging spotted owl can travel without being disrupted by nutrient-insufficient in-dustrial forests. Helicopter logging requires landing pads; trees marked for "landing-pad clearance" were located throughout the section. As these trees tended to be the oldest and largest in the area, their being marked seemed a thinly veiled justification for mini-clear-cuts.

A blue stripe around the circumference of a tree announced its designa-tion for cutting. At one point, I joined Alan and Tess for about an hour in an effort to hand-chip and scrape the blue stripe off of a particularly magnifi-cent tree in the hope that it would be saved. Tess ironically noted that if the "Freddies"[15] were to catch her erasing the blue paint, she could be arrested for damaging a tree. Alan acknowledged her joking by referring to Forest Service staff as "swiney bastards." After the attempt to save the giant Doug-las fir, Tess had the three of us join hands around its trunk, even though it was much too large to accommodate our collective reach. A prayerful mo-ment of silence ensued. I was uncomfortable with my participation in what struck me as an awkwardly manufactured New-Age ritual. My discomfort was not over my hope for the tree's survival but, rather, over engaging in an act that I would not have undertaken on my own.

During the trek back to camp, Tess and Alan wondered aloud how people could cut down these extraordinary trees. What kind of people might they be? ("Who could do such a thing?" "Imagine wanting to cut this down!") I knew the people who toppled these trees and they seemed equally likeable, equally understandable. The echo of the opposition's voice, no matter which group I happened to be with, seemed always to haunt me.

Return to camp was distinguished by a visit from a local Forest Service ranger. Ostensibly, he was there to warn us about the commencement of deer hunting season. But his remark, "There are hundreds of hunters with guns and we can't protect you," was clearly counsel to leave. It was a "you'll-get-what-you-deserve" message, and the ranger's admonition was eerily prescient. Just after I had settled into the dampness of my tent (about midnight), several gunshots rang out, followed by the muddied spin of truck tires. An anxious scurrying and yelling ensued. Dressing quickly, I felt my way down to the group convening at the campfire. Someone from our camp, whose tent was closer to the road, had heard a male voice yell, "Aim for the lights, man!" No doubt he'd meant campfire lights, as the night was otherwise moonless and dark. Another activist, who had been sleeping in his roadside-parked van, said he'd counted just over fourteen rounds. His descriptions of the "two men in a red pick-up" seemed aimed at conjuring up beer-drinking hunters. Fortunately, everyone was safe.

I sought comfort in the fire's mesmerizing glow – "hippie TV" to the good-humoured man on my left. Officially we were awaiting word from those who'd gone into town to report the incident. One by one those who had been asleep returned to the campfire for an emphatic narration of events. I was touched by a local woman's response to the wide-eyed, "why" inquiries of her pre-school daughter. The mother said, calmly and simply, "They're just scared, like us." There was no talk of revenge, no resentment towards the perpetrators, and no one appeared particularly surprised. A late-night camaraderie settled over us all.

The next day was comparatively calm, though peppered with talk of the previous night. The main occurrences were a strategy meeting and a confrontation between a camp member, Brian, and a logger. Both were from a nearby town.[16] Brian was in his early fifties; he had long hair and wire-rimmed glasses. I later learned that he'd earned a forestry degree from an eastern university but, during the last ten years, his attention had turned increasingly to ancient forest protection. The logger (whose name I did not learn) appeared to be about forty. He also had long hair but wore a billed cap and a multipocketed hunting vest. Unlike Brian, he had very few front teeth. Both men were thin and agile. Brian wore name-brand athletic shoes, the logger wore the footwear of his trade – caulk (pronounced "cork") boots.[17]

When I joined the small group observing the exchange, the logger was leaning on the door of a muddy jeep with immense tires. The bumper sticker on the back read: "Looking for your CAT? Look under my tires." The apparent cat/Caterpillar double entendre seemed to go well with the logger's understandably disgruntled mood. He was angry, he said, about the recent loss of his job, anxious about those dependent upon him for food and housing, and sick of the attempts to "shut down the forests." He complained bitterly of earning less than $400 per month on unemployment insurance.

James, a much younger activist who'd counted fourteen rounds and seen the red pickup the night before, entered the fray. In his early twenties, James had long brown hair (which he combed and swished over his shoulders at regular intervals), perfect teeth, and, atypically (for this gathering), he sported a stylish, weather-beaten leather jacket. James drove an expensive late-model van and had brought with him an expensive mountain bike. He was very good at becoming the centre of attention.

James began to call the logger "bro," attempting to construct a fraternal bond that did not appear to exist. He said that he understood the logger's situation, as he too earned only $350 per month working on an organic farm. (I wondered, of course, about the source of James's material assets.) James was no more the logger's "bro" than was I; the logger, for his part, looked as though he wanted to spit on him. Brian, in the meantime, was listening carefully to everything, looking anxiously from James to the logger. Eager to avoid any further escalation of tempers, he manoeuvred the conversation around to this particular timber sale and its ecological consequences. The logger accused Brian of having never been in the forest. Brian met the challenge by proposing a hike to discuss the sale point by point. They could both talk forests; after all, they had both been in the business. The tension eased as the two men set off uphill. Later, Brian said that he doubted that he had converted the logger but that he was happy to have averted a fight.

The subsequent strategy meeting opened with Tess's explanation of how to communicate in a cooperative and egalitarian manner. No one was to be interrupted while speaking. Repetition (the need to say "I like what ___ said") was avoided by "twinkling" (the raising of both hands, palms out, and the silent wriggling of one's fingers) to express support/agreement. Thumbs-up or thumbs-down gestures were more emphatic demonstrations of one's agreement/disagreement with a speaker. A "block" could be employed (a request to speak against the person with the floor) only when serious disagreements arose.

Strategizing centred on the search for volunteers who were willing to remain at the camp until the cutting began. There were requests for tree-sitters and/or road-blockers. Potential press releases were also discussed, but no one knew exactly when the cutting would begin. Two particular exchanges attracted my attention. The first involved distinguishing radical versus moderate activists within the group; the second involved a brief dialogue between local and urban activists.

The first exchange began with one woman's desire to know "who exactly was willing to get arrested." Admittedly, I lowered my head to avoid eye contact. The woman's call for activist loyalty created a stir in the crowd. Low-volume muttering erupted – the mumbling of excuses for lowered hands. Most were drowned out. I caught only one man's oblique reference to his (il)legal history in another state. He needed to "lay low, stay clean." His companions nodded gravely in support. It was intimated that he had "done his time." In the end, slightly less than half the circle of attendees raised their hands. Once again, Tess assumed unofficial control, assuring everyone that only those comfortable with arrest need put themselves in jeopardy: "Other kinds of work needed doing."

The second exchange occurred during the planning of press releases. I noted that only those activists who lived in the area concerned themselves with the plight of loggers. They emphasized the need to communicate their empathy for loggers "because these guys are our neighbours." This drew some support (thumbs up) and some resistance (thumbs down), the latter coming from the younger members of the crowd. "Yes of course," one woman said quickly, dismissing the concern, "but we must focus on this mountain, this sale."

Two full weeks after I'd returned home, an article covering the protest appeared in the local paper. The story appeared under the headline: "Shots Fired at Forest Activists." The accompanying photograph showed an articulate, soft-spoken activist named Danny. He had maintained a low profile all that weekend, and I can only assume that his photograph appeared in the paper because he was the most hirsute among us. Certainly, his photograph was not representative of the appearance of most people at the gathering. In 1996, despite continued and expanded public pressure, the site's magnificent trees were felled after all legal appeals failed to stop the cut.

Spirit of the Earth Activist Conference, 1993

- A person is rich in proportion to that which they can live without. (Thoreau)
- You may not like living with us now, but conservationists make great ancestors. (Matthew Waite)
- We don't inherit the Earth from our ancestors, we borrow it from our children.
- Hope has two beautiful daughters, their names are Anger & Courage; anger at the way things are, and courage to make sure that they change. (St. Augustine)
- A conservationist's motto: the first rule of intelligent tinkering is to keep all the parts. (Aldo Leopold)
- We must be compassionate and fierce at once. (Terry Tempest Williams)
- By saving the forests we help save the wilds. By saving the wilds we help save civilization. By saving civilization we save the planet, we save ourselves, and future generations.

These were the slogans (most attributed) that adorned the walls of the principal meeting hall at the 1993 Spirit of the Earth Activist Conference convened annually in a small college and retirement town in rural Oregon. It is the premier event for grassroots environmentalists west of the Cascade Mountains. Ironically, the building in which it takes place is part of a former military installation. During the year, the facility's smaller rooms function as headquarters for a variety of activist groups. This inexpensive conference is supported by private donations and meagre attendance fees. Those who cannot afford accommodation are billeted in local homes. I stayed with a lovely woman and her eccentric teenage son, neither of whom was otherwise involved in the conference. She simply left her back door unlocked and insisted that I make myself comfortable.

The mood during the opening presentations was reminiscent of a revival meeting. A welcome from event organizers was followed by a prayer and song that were delivered by an elderly Takielma woman and a Karuk man, respectively. Their melodic invocations to "speak for the voiceless inhabitants of the forest" commanded silence in the crowded room. A majority in the audience answered these opening prayers with raised arms and shouted: "Ho" – a gesture common in New Age and men's movement circles.

For three days, the impassioned words of panelists emanated from the plant-adorned stage. Opposite the stage, at the far end of the hall, activists milled about embracing old friends, talking animatedly, and making their way past a makeshift set of booths. Posters, T-shirts, brochures, and "planet-friendly" products were available for consumption. Participants' clothing ranged from paisley and tie-die to pearl-buttoned plaid shirts and cowboy hats.

Most lectures and panels followed one of three official goals: (1) to form an alliance between organized labour and the environmental community; (2) to conduct legal and political brainstorming aimed at influencing the Clinton administration; and, especially, (3) to communicate scientific information.[18] Talk pointed to who among President Clinton's current advisors was thought to be "green," to progress made (or not) in specific court cases, and to establishing the coho salmon as the next northern spotted owl. But the events that attracted the most attention were: an activist's personal testimony on grief and spirituality, a logger's speech, and an impromptu women's protest.

An Environmental Activist Speaks

Lila Mountain has been an environmental activist for more than twenty years. She was supposed to speak on jobs and on restoring watersheds and fish habitat, but she soon abandoned that agenda. Instead of standing at the microphone, she remained slumped in her chair with a shawl wrapped around her upper body, affecting the wisdom of an aged crone. A quieted audience shuffled forward, closing in on the limited range of audibility. "I have decided," Lila began, "on a personal statement as I have been in a funk," troubled by depression, and haunted by a "heaviness of heart." This stemmed, she speculated, from knowing "so little about the connectedness of the Earth" from "not having a clue." "There is no way," she concluded, "that we can ever put Humpty Dumpty [i.e., fractured ecosystems] together again."

The problem [is that] we've lost our native intelligence, the wisdom in the cells of animals ... and so when we approach ecological problems ... we too often bring to [them] compartmentalized thinking ... There are no techno-fixes. We need to restore our relationship to the natural world ... [The] problem [is the] broken bond between the human community and the natural world ... Obviously we need an economic revolution to a highly informed, hands-on culture ... [we] need to evolve a cultural ecology ... to develop models for resource stewardship.

The audience appeared to be deeply moved by Lila's solemn invocation of cultural and economic revolution, by her sense of futility regarding the repair of the watershed and its spawning bed. She was granted a sustained standing ovation, and her speech was endorsed throughout the weekend. All of this surprised me, as I found her language clichéd and the apparent heaviness of her manner a direct contradiction to the conference's otherwise exuberant mood. Perhaps the frequent references to her address were a means of reminding the audience that the matter at hand – ecological survival – was not to be taken lightly. Confidence and exuberance were to be checked by the gravity of the activists' goals for the Earth.

A Logger Speaks

Mark Evans is an independent Oregon logger and activist with the pro-timber Oregon Forest Community Coalition. He is lean and self-assured, wears frameless glasses, and has a presence more suited to academia than to logging. He is also an articulate speaker – known, admired, and widely acknowledged among fellow loggers. The invitation extended to Evans was controversial, though it is indicative of organizer Andrea Smithers' expansive personality.[19] Evans surely knew that the success of his presentation depended upon his ability to touch the resistant hearts and minds of his audience. So he began by drawing an analogy between himself, a socially "vilified logger," and the other prominent political (and activist) dispute of the period, Oregon's narrowly defeated Ballot Measure 9. The draconian measure introduced by Christian fundamentalists was designed to permanently bar Oregon's gay and lesbian population from legally sanctioned equal rights and, in some cases, public employment. It too had widened the gulf between urban and rural Oregonians.[20] As Mark began, he, like Lila, remained seated, shoulders drooped and head cocked away from the audience. His posture was confessional, almost intimate, his voice low. Unlike his wife and rural neighbours, Mark said, he endorsed gay rights because he knew "how it felt to be persecuted." This was followed by an appeal to support "the small business side of logging" and his insistence that he was not the one "to put out of business." And then, with raised shoulders and palms turned outward, he said: "I love being a logger; it teaches me humility;

when you think you've learned it all, you've learned that you know nothing at all." "Nothing," he said, would please him more than to "create an alliance between small-time loggers and environmentalists."

Attending gay-supportive activists crowded the stage stairs as the panel of which Evans was a part concluded. Evans received the only other standing ovation given a speaker that weekend. Did attending activists, I wondered, need a logger with whom they could identify to make them realize that what they formally opposed was the logging industry and not loggers per se? If so, then they found this person in Mark Evans. Others, however, resented his presence. An environmental law student with whom I had lunch that day accused him of being "emotionally manipulative," while a subsequent speaker sarcastically apologized for not possessing "Mark Evans's mastery of dramatic pause."

Women Protest

A hand-written sign at the front door greeted conferees on the morning of the gathering's last day. It read: "Tired of being invisible? Meet for a special session on 'Sexism in the Environmental Movement.'" This session, which was not part of the established agenda, was organized by women who rightly contested the lack of female speakers at the conference. Despite the attention given to Lila, the two preceding days were conspicuous for the repeated on-stage presence of a core group of well-known male activists.

When I entered the small meeting room, it was already filled to capacity with about thirty women and five men. Tensions were high, as participants debriefed the conference's distinctly masculine atmosphere. Some women wanted to storm the stage, rearrange the format of the remaining day's events, and request a formal apology from the movement's established male leaders. Others were nervous about creating divisions within the movement as a whole. The altercation culminated with an exchange between a woman who insisted that the group collectively invoke the calm of "women's traditional role as healers" and a woman who asserted her right to "express anger." The latter woman insisted that the problem was one of male activists' "unevolvedness"; that is, she believed that men were insufficiently conscious of gender politics and had thus failed to engage women in the movement. The first (and only) African-American woman, a Californian I had encountered in my Oregon-activist travels, then tactfully silenced the "evolved" discussion by critiquing its monolithic standard for behaviour.

Some part of the women's anger was, in fact, sparked by a "joke" offered from the previous day's front stage. Celebrated activist Matthew Waite had jested about his (mostly male) stage-mates competing over the size of their cell phones because they were somehow equated with the size of their (male) anatomical endowments. Several viewed the joke as offensive and/or inappropriate. Ultimately, the incentive it provided for the meeting was productive. It was decided that four women would present a cursory set of guidelines to help achieve equal representation at future meetings. A draft set of recommendations emerged in the months that followed; these "Conference Principles" became, effectively, an affirmative action policy for the movement. The central features of this policy provided an important template for discussions of inner-group recognition and participation specific to Oregon's ancient forest movement.

It is difficult to summarize all that occurs in the myriad events that constitute a conference. Certainly, I was impressed by how these activists eagerly consumed reports from field scientists – be they concerning factors contributing to the maintenance of anadromous fish runs or the latest information on mycorrhizal fungi. If, at times, I tired of the gathering's rhetorical style, the problem was my own. Conferences offer essential opportunities to communicate information and to recharge one's activist batteries. A component of this is the relief provided by the chance to drink from a pool of common language. It means that one need not, for a few communal days, explain oneself when using the phrase "Mother Earth," "biotic community," or "endangered habitat." The annual grassroots conferences are, writes Kathie Durbin (1996, 189), "tribal affairs where tree huggers can let down their guard ... and reinforce each other in the work they all share."

Old-growth forest in western Oregon. Note the decaying, fallen logs so important to the forest's productivity and biomass.

Photographs by Elizabeth Feryl

The machines in the garden. Note especially the tower at rear, which is a smaller version of the tower used for hauling logs described on page 44.

Aerial view of the patchwork pattern of clear-cuts and young second-growth forests in the Gifford Pinchot National Forest just north of the Washington-Oregon border.

62

Young, replanted second-growth forest in the Willamette National Forest, Oregon.

Clear-cuts on private land near Eugene, Oregon. Note the web-pattern scarring from the tower cables, as described on page 44.

4

Negotiating Agency and the Quest for Grassroots Legitimacy

Democracy, and the increasingly competitive voices of civil society, is – to put it bluntly – a wondrously messy affair. This is in part because the revered and ever-changing institution must somehow accommodate the inclusive principle that all citizens have rights worthy of consideration within the political arena. Rights are not solely the indispensable but dry points of law that permit one to vote in the next election. "They are expressions of our moral identity," including the deep sense of justice that propels group-centric pursuits pertaining to the right to language, Aboriginal title, or protection from the sometimes less than egalitarian will of elected majorities. And, ironically, there is no better vehicle for "denying the rights of others than to claim that others are denying your own" (Ignatieff 2000, 1-6).

Chapter 3's examination of the worlds of Oregon's loggers and environmental activists hints at the primacy of rights. In the quotidian and overtly political movements of activists, one can see that participants in both struggles assert their right to live in and imagine a cultural world drawn in their own image. The undeclared goal of their duel is to have one or the other of the identities being promoted become the prototype (more or less) for a reshaped forest policy. That very personal and local struggle (however significant its larger ramifications) constitutes the day-to-day lives of many of Oregon's grassroots activists.

However, the idea that loggers have rights was relatively new in early and mid-1990s Oregon. Historically, environmentalists were the only forest-preoccupied grassroots players; they were used to competing with large, abstract forces – Congress, the Forest Service, corporate power, and other socio-structural expressions of late-twentieth-century capitalism. But the emergence of the lands movement in the American west and, particularly, worker- and community-driven entities such as the Oregon Forest Community Coalition presented environmentalists with a new problem. Activism, forest activism, was no longer solely associated with liberal politics; rather, environmentalists were forced to face the momentum of rural (often conservative) grassroots citizens interested in the utilitarian appropriation of public land for logging.

A close examination of the contest for grassroots power (relatively speaking), which is manifest when one compares environmentalists' descriptions of timber workers with the latter's descriptions of themselves, illuminates the hard work of achieving and maintaining grassroots status. Historically, environmentalists had something of an upper hand because of their many prior decades of organizing and because of the advantage conferred by widespread public support for environmental causes (Dunlap and Scarce 1991; Ladd and Bowman 1995).[1] The victory of President Clinton's new forest plan was also pertinent as the plan pledged to reduce timber harvests to one-fifth of their late 1980s levels (FEMAT 1993). But advantage in politics is invariably fleeting, and so environmentalists struggled to maintain it by tacitly ignoring the growth of timber activism and by portraying loggers as instruments of "big timber."[2] Most clung to the view that timber activism was "little more than a collection of phony citizen groups organised by big business to do its bidding" rather than a growing movement in its own right (Brick 1995, 19). This tendency to dismiss timber workers as pawns – passive instruments under the control of corporate timber – was paired with the framing of rural logging communities as decaying, diseased, or dying; that is, as pathological.

Timber workers, alternatively, spoke in defensive terms as they sought to promote their legitimacy as grassroots actors. They embarked upon a discursive struggle to be taken seriously, to be seen as suffering individuals subsumed under, yet distinct from, "big" timber's reputation for corporate greed and environmental devastation. In other words, timber workers resisted environmentalists' characterizations of them by regularly invoking the injustice of economic hardship and, especially, their experiences of social stigmatization. These invocations were essential to the formation of a politically effective group identity. They helped establish the public presence of a new grassroots force; namely, "timber folk" capable of action, deserving of rights, and worthy of sympathetic attention. Stigma, therefore, functioned as a discursive tool for the invocation of political recognition (Taylor 1992) and the formation of politically effective solidarities.

It is to this discursive, identity-constructing struggle for grassroots status that this chapter turns. The point is not to determine which grassroots contingent played "fair" and which did not; rather, it is to look at what happens between opposing grassroots players when the pursuit of capital and the promotion of profit has already depleted the forest resources in question at the expense of both (Foster 1993, 4).

Timber Worker as Pawn

Matthew Waite has dedicated more than two decades of his adult life to protecting ancient forests. A pugnacious political strategist with Ancient Forest Grassroots Alliance, Waite is apt to measure his worth in terms of those who dislike him. On more than one occasion he, like Ralph Nader, has paraphrased President Grover Cleveland, claiming to be "deeply loved for the enemies he's made." Much to the chagrin of timber advocates, the media never tired of Waite's pregnant soundbites. His phrases appeared everywhere, taking on a life of their own: "Environmentalists may be hell to live with, [but] we make great ancestors"; "Expecting the Oregon delegation to behave rationally about the end of forest cutting in the 1990s is like expecting the Mississippi delegation to behave rationally about segregation in the 1960s."[3] I'd long tracked Waite's activities in Oregon, though my first formal interview with him took place on the day President Clinton announced the draft management plan for northern spotted owl habitat (Thomas et al. 1993).

During the interview, one primary metaphor dominated Waite's conception of timber workers: the timber worker as pawn. The image speaks literally of pawns on the chessboard, the smallest of the players under the command of industry's all-powerful hands. It portrays workers as victims of those who possess ideological and material control of production. Matthew Waite provided a salient illustration of the logger as pawn when speaking to a civic organization earlier that year. The illustration is accompanied by an overture about the deep resentment timber workers are said to feel towards environmentalists, a resentment Waite diagnosed as an outgrowth of loggers' social and political impotence:

I think environmentalists and timber workers have the same long-term interests – sustainable forests. But most workers don't see it that way right now. They identify more with owners and management, not the environmentalists ... Many of these workers are ignorant. These workers are powerless. They've been jacked around by politicians, and, they perceive, the environmentalists.

When intimating that the attempt of timber workers to organize and empower themselves at a grassroots level is the brainchild of industry puppeteers, Waite is effectively defining loggers as passive puppets and, thereby, negating the possibility of their being involved in any form of self-directed grassroots activism. "In a twisted, and, I think, cruel sense, management has in a token way empowered the workers for the first time. Basically management said to their workers: 'You know, if you throw a tantrum about the spotted owl, if you get upset, if you organize, it will go away.'"

Marie Jarmon, a self-described activist-ecologist and teacher whom I interviewed, has travelled across the country with her educational slide show on ancient forests. When we met, she impressed me with her judicious manner and willingness to work as an educator in timber communities despite the fact that on one occasion her car was vandalized, its tires slashed, and her "green" bumper stickers removed. Her teaching included guided hikes with male high school students – the next generation of loggers – which she undertook in order to explain to them the non-timber value of forests. Though compassionate, Marie believed that, for the most part, loggers "didn't know any better" and that, once "educated," they would likely see the world as she did.

Marie also shared Matthew Waite's views of timber workers as pawns and incorporated that vision into her ideal solution to the forest dispute. For Marie, loggers' lack of consciousness regarding their exploitation is a considerable source of frustration as well as a barrier to any possible resolution to the forest conflict.

> The trouble is these big money interests ... They're the ones that are profiting off all of us. You know, they get the workers to fight with the environmentalists. In the meantime, they're running off with the bucks. That's what I was really trying to get people to see. If the workers could see that, [they] could see that they're being exploited.

The substantial historical precedent for Marie's claims should not be ignored. The destruction of communities and the depletion of resources have distinguished the political economy of Washington, Oregon, and northern California (Robbins 1989; White 1980, 1992, 1991). Nonetheless, from the standpoint of portraits of identity and agency, the above exchanges serve additional ends. When environmentalists consistently characterize their logger opponents as pawns, they are implying that they (environmentalists) are more insightful, more conscious than are loggers and, therefore, that they understand loggers and their situation better than do loggers themselves. This characterization defines loggers and mill workers as incapable of taking action or of having opinions of their own. Rhetorically, it also denies the existence of a pro-timber grassroots movement with its own base of support – a base of support that is community- not industry-powered. Such ploys serve to de-emphasize loggers' claims to belong to a community-based, timber-advocating activist movement, as well as their attempt to construct an identity based on their status as timber workers embedded in the social fabric of rural Oregon. Pawn accusations also reflect what some social movement analysts refer to as collective action frames: interpretation schemas that help attribute meaning, define reality, and guide audience perceptions of events (Snow et al. 1986; Snow and Benford 1992).

Interestingly, environmentalists did not always draw a clear line between loggers as pawns and environmentalists as conscious agents (i.e., anti-pawns). When invoking the pawn metaphor, some would periodically include themselves in the characterization; that is, they were often quick to add that they, too, were victims of corporate timber's actions. This implied the existence of a shared community, a world of mutual victimization. The following examples suggest why this depiction of inherent mutuality makes sense.

I asked Andrew Simon, a southern Oregon environmentalist, where he placed loggers in the forest conflict. He called into play the same corporate history referred to by Jarmon; however, after a brief pause, he placed environmentalists in the victim pool. "Victims, victims, they're pawns [pause]. But we're pawns too. In other words, essentially what's going on is a much larger issue, and the issue is really the fight between large organizations and small organizations." To his credit, Simon acknowledged the role urban-generated, progressive movements have had in representing rural peoples as "uneducated dolts." But he placed most of the blame for pitting "owls and uncaring environmentalists" against loggers on the timber industry's symbolic manipulation of the logger. Praising the industry for its deviousness, Simon offers an argument that parallels the work of historian Peter Walls (cited in Chapter 2). Like Walls, Simon credits big timber for merging the image of the logger with the image of the cowboy, arguing that industry's public relations machine has constructed a new American hero by romanticizing this category of manual labour.

Undermining Cultural Romance

Romantic constructions of loggers and logging communities were frequently condemned by environmentalists, who referred to the timber industry and timber communities in terms of disease, dysfunction, and death. Timber communities, in particular, were represented as plagued by physical and psychological ill-health. It follows, of course, that if a world is diseased, then so too are its inhabitants. Ironically, the "diseased" attributes that environmentalists tended to project onto timber communities mirror the descriptions of trees and forests that one finds in the writings of early White pioneers and travellers. "The trees ... were too big, too dark, too cold for the brave new man. The trees were 'indecently large,' unheard of, improper. Visitors complained of a 'peculiar melancholy,' of trees 'startling and strange' ... This 'interminable, apparently impenetrable, thicket of firs exercised upon my mind, I confess, a gloomy, depressing influence'" (Tisdale 1992, 42-44). This same sense of doom and melancholy appeared in environmentalists' descriptions of timber communities. Sundry references were made to the grim nature of timber work: the logger's long hours and his (and, very rarely, her) constant vulnerability to physical danger. Loggers are depicted as doing work that no being of sound mind and body would rationally endure, and as the only remaining inhabitants of afflicted and perishing communities. In Andrew Simon's words:

In terms of public perception, they've been turning the logger into the cowboy. And the relationship between what a logger is really about, ... you know, compared with that image, is about the same ... as what it is to be a cowboy. It ain't so much fun. You're talking about a profession or a trade in which one in six are seriously maimed or killed. You're talking about highly, to use modern terms, dysfunctional communities in which the statistics of social pathologies are very high, and alcoholism and everything like that. And I'm not singling it out, but what I'm saying is, this ain't a great life ... Is this the valuable piece of Americana that we're trying to save as some indication of the good life? I mean, nonsense.

Matthew Waite, the acerbic AFGA spokesperson introduced above, was not a stranger to frequent verbal assaults from timber advocates. Eerie cloth effigies of his bearded face and stocky torso were a regular and early feature at timber rallies. Matthew held his own, however, by relying heavily on pathologically charged analogies. One of his favourites involved the claim that timber workers are like drug addicts. He had heard, he would say, the plaintive cry of loggers hurt by their portrayal as pariahs, by the tacit reminders that their profession is "no longer honourable." But Matthew insisted that, due to their dependence upon publicly owned forests, timber communities are like addicts: they are "addicted to timber dollars."

Waite's characterization is shared by others. Carolyn Armstrong, an AFGA activist, views this "addiction" as a significant obstacle to finding a cooperatively negotiated solution to the forest dispute.

The trouble is ... I mean I also see this as a, like an addiction thing – you know, that cutting down the trees is an addiction. And that the people that are in that culture of logging the old growth are in denial of the fact that it can't go on forever. I mean, I watched this show, *Rage Over Trees,* and here's this guy, I forget his name, the guy that owned a mill or something, and he kept repeating over and over again, "We're never going to run out of old growth." It was like, what is wrong with this guy? Talk about denial, you know. So ... in that sense, you know, when somebody's addicted, you can't really reason with them. That's when I tend to believe that you've just got to stop them with a lawsuit or something.

Ultimately, these characterizations recast timber workers from natural resource labourers, whose end products are widely coveted on local and global markets, into workers responsible for their plight because they have chosen their occupation and, subsequently, become "addicted" to it. Further, it is implied that "denial" is all that stands in the way of a liberation that can only be understood as resulting in "freedom" from a livelihood based on the cutting and hauling of timber.

The coupling of sympathy for pawns with an attribution of psychological ill-health was beautifully captured in Matthew Waite's above-noted community lecture. Waite consistently recognized the suffering of timber workers, as opposed to owners, and depicted them as victims of changing economic times, industrial demands, and environmental priorities. His explanations typically and appropriately included structural variables: macroeconomic forces, log exports, manipulation at the hands of industry, and so forth. And he rightly insisted that there are "no easy answers," a sentiment with which many would agree. But his message was mixed. There was an uneasy tension in Waite's speech, a tension born of the desire to project, simultaneously, sympathy (acknowledgment of the suffering of timber workers) and aversion ("they're like drug addicts"). Thus, when he asked himself what could be done to *help* victimized timber workers, Waite called on Elizabeth Kubler-Ross's (1969) now famous stages regarding the acceptance of death in order to account for the behaviour of members of timber-dependent communities.

> Psychologists have defined several distinct stages of dealing with the death of a loved one or loved thing. That first stage being denial, and we hear that with a quote saying: "Spotted owls don't need old growth forests." The second stage is anger, and I think that is represented by the cliché, "Spotted owl tastes a lot like chicken" (actually I hear it tastes a lot more like bald eagle, but ... I haven't ...). The third stage is bargaining, and this is, "We can have the spotted owl and log the old-growth forests too." The fourth stage is grieving, and they say, "Environmentalists are practising cultural genocide." The fifth stage is acceptance, and we've gone through a lot of the stages, but not all have accepted it yet.

A clear implication of Waite's model is that he also views timber communities as dead or dying. Of equal importance is the overall attribution of psychological ill-health: one can virtually see timber communities lying on the psychiatrist's couch.

Reconciling Ambiguity

How, then, can we make sense of environmentalists' mixed characterizations of timber workers? Many environmentalists express sympathy for loggers and yet also move to limit their political power by accusing them of ignorance and passivity. The crux of the matter is that environmentalists' characterizations of timber workers are intrinsically ambiguous. Identity and discourse theorists have proposed ideas that are relevant within this context. Lincoln (1989), following Durkheim (1965), has argued, for instance, that discourse (aided by its capacity to evoke sentiment) is regularly employed to emphasize the links that join people together, to remind them of their loyalties to one another. Identity theorists, meanwhile, argue that social movements are not groupings fixed in place at any one point in time; rather, they are informal, fluid networks of people who, through their evoked commonalities, come together as a group and feel bound by their collective sense of belonging and their shared vision of the world. As a corollary to this, they distinguish themselves as separate from those who fall outside their group (della Porta and Diani 1999, 17; Holland et al. 1998, 19-32; Lincoln 1989, 9). Thus, on the one hand, certain characterizations of timber workers can be construed as appeals for group formation (we are all victims of timber corporations), while, on the other hand, some characterizations can be construed as appeals for distance (timber workers are passive pawns, psychologically impaired; they're like drug addicts, addicted to timber dollars; their distress is self-inflicted).

The confusion lies in the fact that this period of research was marked by social and political transitions. A shift in national policy regarding publicly-owned forests had occurred. During periods of transition and political struggle, opposing grassroots groups resemble players at either end of a teeter-totter whose movement is deeply affected by state, or macrolevel, political decisions. During the transition period, environmentalists, as noted, had the upper hand. The ambiguity inherent in their characterizations of timber workers was connected to that upper hand and the transition generally. Environmentalists needed, at one and the same time, to balance a grassroots, anti-hegemonic (i.e., "underdog") role with a superordinate role. They struggled to emerge as the premier grassroots group while simultaneously expressing their sympathy for timber workers and controlling the public perception of the latter's grassroots organizations. In short, they were faced with a difficult task that required both the strength of Goliath and the persona of David.

Loggers: Stigma and Group Identity

Having established that environmentalists communicated pejorative and/or ambiguous messages about timber workers, one must ask what resemblance these characterizations bear to the image that members of timber-dependent communities hold of themselves. Does their group self-portrait converse with the one put forth by environmentalists?

The most salient point here is that many timber workers have come to view themselves as socially stigmatized within the overall climate that encompasses the forest dispute. Stigmatization refers to a "mark borne by an individual or group that causes that person (or persons) to be socially discredited ... [and thus] reduced in our minds from a whole and usual person to a tainted, discounted one" (Goffman 1963, 1-3). Stigmatization, as an experience, was first raised in Chapter 3 with reference to veteran logger, Jim Stratton, whose worksite I visited. As noted then, Jim had attempted to combat derogatory attitudes towards loggers by writing a letter to a local newspaper. In that letter, he told a story about a conversation in a line-up at a fast-food outlet. "One of the businessmen paused and said, 'Say, what's the hold up in here, anyway?' His companion replied, with a derisive laugh, 'I don't know. Maybe they're still retraining one of those dumb loggers.'" The letter went on to document representations of loggers as "drunken, piggish louts whacking down trees with reckless abandon, rubbing chain saws in glee at the destruction of the forest and its wildlife." These projections of obtuseness, delight with destructiveness, and the notion of displaced loggers exiled to poorly paid work all adhere to the kind of marginalization said to be a by-product of the stigma theories we use to rationalize our animosity (Goffman 1963; Jones et al. 1984). In this story a businessman equated the restaurant's poor service with stupidity – the epitome of which is embodied in the logger.

Timber advocates frequently paired references to their stigmatization with references that portrayed their lack of access to public forums – a lack they believed increased their victimization. Lorna Owen is a middle-aged woman with grown children who had come to devote much of her time and energy to Save Our Community (SOC). Lorna is a bookkeeper for Mason Logging, which operates out of a small white clapboard house on one of the main streets in Caledon, Oregon.[4] This "headquarters" for the Mason family's small logging operation is also the physical home and nerve centre for SOC, which identifies itself as one of the many grassroots subsets of the Oregon Forest Community Coalition. I spoke with Lorna on three different occasions.

Lorna's perception of her group as victimized was primarily due to her feeling that the economic plight of timber workers was of little concern to the outside world. She told me of her response to a reporter's inquiry about the release of thirteen Bureau of Land Management timber sales. The release of these sales, notes Lorna, will do nothing to relieve the distress of timber workers. Thus, she responded angrily to the reporter: "Does that make you feel better? I mean you've lost your home you're living in, you're living under a bridge, and you can't feed your kids, and you have no medical insurance, but you know maybe three or five years down the road, maybe you can get back to work." Lorna insists that her community's story won't make it to the papers. The *real* story would inform the public that released timber sales are not synonymous with immediate logging employment. Lorna described the reporter as interested and sympathetic, but, in the end, her story was not printed.

The depiction of timber workers as "disposable" was accompanied by a second story about Lorna's attempts to gain access to a locally televised town meeting that followed the showing of an Audubon Society documentary film on the forest conflict. When the scheduled event arrived, Lorna was not able to find the producer with whom she had been communicating.

The new producer denied her access. In comparison, Lorna claimed that environmentalists who didn't have reservations had free access to the event. Moreover, she insisted that those from her community who did gain access had to do so by posing as environmentalists. "One of the secretaries from BCL [the local mill] ... when we go in and see her, I said, 'How in the world did you ever get in here?' and she said, 'I told them I was for the owl.' But ... we were wearing yellow ribbons."[5] At this point in the conversation Lorna's talk of denied access trailed off; talk turned instead to hecklers planted in the audience by the TV station and the presence of police officers. Lorna intimated that people thought "we [loggers] were going to destroy everything."

Lorna's comments communicate several messages. Obviously, she senses that her community's "voice" is frequently silenced, and that the urgency of the plight of timber workers is largely invisible. She believes that loggers are perceived as inherently "destructive," and that members of timber-dependent communities are punished or socially ostracized for holding a distinct set of moral standards. Lorna's experience tells her that the ability to expose one's "voice" depends upon one's proximity to environmentalist points of view. Implicitly, Lorna's rhetoric also serves to define herself and her group (SOC) as defending the economically impoverished, not the industry per se. Finally, Lorna's comments about exclusion from the forum on the basis of group identity defined the conflict as sociocultural rather than environmental.

Lorna's hyperbole is obvious, but the "truth" of these accounts, as she told them to me, is not at issue here. What matters is that Lorna has come to feel marginalized, stigmatized. Her concerns invoke democratic beliefs about identity and fairness. A "fair break," in Lorna's case, refers to her desire for public representation (a voice) and relief from the burden of attributions of moral corruption based solely on one's ideological support for, and residence in, a timber-dependent community.

Goffman both influenced and was influenced by psychological paradigms (Bock 1988); this influence is particularly apparent in his portraits of social actors encumbered by feelings of anxiety and/or social insecurity. In *Stigma,* Goffman explains the tendency of the stigmatized to become socially isolated and, therefore, suspicious, depressed, hostile, anxious, and/or bewildered. However, what is both implicit and explicit here is the desire for psychological relief, a desire born of having once known a different social climate, hence, a different (i.e., non-stigmatized) social status.

Laurie Willings and her husband own a two-employee salvage logging company. They focus on the clearing of small parcels of land: "No job is too small." Laurie was initially reluctant to talk with me, an uncommon occurrence in my experiences in timber communities, but she was persuaded to do so by another activist whom I'd interviewed. Laurie is articulate, witty, once edited Caledon's local paper, and was an active member of SOC. She became involved in the old-growth dispute early enough to have others suggest that she was overreacting. Nonetheless, she sought answers to her own questions: "What's the deal with the spotted owl? Are we really out of old growth? Are we really doing this to our land? You know, I would ask these questions because I wanted to know, is this really true, are we these bad guys?"

In answering these questions with a resounding *"No,"* Laurie invoked her husband's experiences as a third-generation logger who "liked to go to work and come home to his family at the end of the day." She described him, and the "many men like him," as "extremely sensitive because they've worked hard, it's the hardest work. And all of a sudden they're bad guys for doing something, for providing a product to a country that demands it." In other words, Laurie contested negative portraits of loggers by defining her husband (a logger) as emotionally sensitive and socially decent (i.e., hard-working). It was not that loggers were by nature sensitive but, rather, that they had become sensitive because their work was no longer socially appreciated.

Laurie's experience of prejudicial treatment also motivated her desire to belong to the world of the socially acceptable. "Sometimes, I don't tell people a lot of times that I'm half owner of a logging company ... I don't want to get trashed for that. Many times I have wanted to close my doors and sell our equipment and start up something that the environmental community would look favourably on."

The stigmatization of loggers is a fairly new phenomenon. Until recently, rural Oregon actively celebrated the role of timber work, associating it with the state's personality. I can find 1950s postcards in local antique stores that prominently display clear-cuts as part of the state's scenic panoramas. I am reminded of this when speaking to an acquaintance who moved to Eugene twenty years prior and now edits a computer magazine. Upon arriving in Oregon, she fell in love with the lush forests and the west coast's relatively untamed coastline. But her retrospective comments reflected a remarkable change in perception over a relatively short period of time. She recalled that, in her early years in Oregon, she used to watch the logging trucks roll down the highway – trucks that then struck her as quaint, "picturesque." Her current impression of those same trucks couldn't be more different. She now sees the logs on the truck's bed as "stacked, dead bodies"; trucks, whose material content changed very little in the intervening years, now conjure up a foreboding, macabre presence.

The shift from romantic, or at least benign, images of logging to highly stigmatized ones is something that members of timber-dependent communities frequently mentioned. This is of particular importance because it both develops and confirms the point that what is stigmatizing in one social context is not necessarily stigmatizing in another (Goffman 1963). The natural history of a category of persons with a stigma must be clearly distinguished from the natural history of the stigma itself – the history of the origins, spread, and decline of the capacity of an attribute to serve as a stigma within a particular society (e.g., divorce in American upper-middle-class society) (32).

The transitive quality of a stigma is of considerable importance to timber workers. Because a historical shift has taken place, experiences of stigmatization are framed in terms of a dual knowledge: the combined self-knowledge of the logger as stigmatized and the logger as "normal."

Inside Caledon's Timberman's Diner the noisy lunchtime crowd mutes the sound of the heavy winter rains against the streetside windows. Beverly Mason (of Mason Logging), Bill Hawkins, and I talked through lunch about their activities with SOC. At one point Bill reiterated portions of my earlier conversation with him; namely, those concerning his feelings of being misunderstood, as being seen by "the other side" as evil and destructive. Beverly picked up this thread and pursued it.

> I think Paul Bunyan is the epitome of that. You went from the lumberjack hero image and now you've been shifted around. You're wearing the black hat. You're the villain. To have their children come home from school that way, feeling that you're doing something you're ashamed of, that your dad is. It's a very difficult thing [for] even the logger – not just the children – but even the logger to deal with. One local example is that it's always been a thing in the Buckaroo Rodeo [Caledon's] to have a parade. That started around 1913. The loggers would always get their trucks all shined up and clean and pretty, and whatever. And then you would go out and get the biggest tree log that you could. The straighter, the larger, the better. And the last few years we don't even enter our trucks, because you're bad, there goes that last old-growth tree. You've become a symbol of evil instead of pride ... You find yourself characterized that way in newspaper articles. One of the cases that we noticed was an assault murder case, and it seemed as though every paragraph said this fellow was a logger. And you wondered if they would have said "Joe Blow accountant" every paragraph.

Of particular note here is the progression of meaning accorded symbols of key importance to timber workers. "Black hat," a reference to similarly clothed villains in western films, is a metaphor that captured the fact that, historically, loggers were identified with the "good/white hat" cowboys of the silver screen. Similarly, Beverly's reference to "bad" loggers and the "last tree" invoked the resentment generated by the accusation that loggers are responsible for denuding the west of trees. Loggers frequently attempted to counteract this accusation by drawing attention to everyone's consumption of wood products, as well as by recalling stories of easterners and urbanites who were "shocked" to find any deep green hillsides remaining in the national forests at all.

Beverly explained the transformation of loggers from heroes to villains via the socially imposed re-costuming of loggers (from white hat to black hat) and via the denigration of cultural icons (i.e., logging trucks). For his part, Bill Hawkins, a second-generation forester and member of SOC, personalized the transformation by narrating the story of a heated exchange with a childhood friend. The friend-turned-foe now saw Bill as a kind of "Darth Vader" because of his endorsement of logging: "These sorts of things just flash back and I realize how deep a hole we're in [with regard to] public perception"; "What shocked me is that, you know, [we're] hard-working people in business [and just] because we cut trees we're [presented as] so bad."

In the end, both Beverly and Bill Hawkins compared past and present attitudes towards logging and loggers. The clear message is that who they are and what they do for a living has stayed the same; it is society that has changed. While Goffman (1963, 34) does not explore these shifts (what he refers to as the natural history of a stigma), he does consider the subjective experience of once "normal" individuals who acquire a stigma (e.g., blindness): "Presumably [this person] will have a special problem in re-identifying himself, and a special likelihood of developing disapproval of self."

While there is some evidence for "disapproval of self," the more obvious evidence shows that stigmatization coexists with the promotion of pride in self due to one's status as a logger, mill worker, or simply as a member of a timber-dependent community. This emerging group identity is not simply a manifestation of shared deprivation – what structural-functionalists once deemed the maladjusted by-products of an otherwise cohesive society – rather, it is a movement aimed at producing new cultural worlds (della Porta and Diani 1999, 4-5; Holland et al. 1998). Indeed, one of the defining features of the forest dispute in particular, and of new social movements in general, is the prevalence of identity-based self-determination (Johnston, Laraña, and Gusfield 1994). Loggers have come to define the conflict as an intrinsically cultural one, and they have come to celebrate, even elaborate, their identity at rallies, conferences, and in their social world generally. Hence, the popularity of the idea of the logger as an "endangered social species," the claim that the Clinton Plan threatens to extinguish a social group unique to the Pacific Northwest, and the ubiquitous posters and bumper stickers calling for protection of the loggers' "heritage." In the words of one timber company employee and songwriter:

What about my family, what about our home?
Wildlife and timber, it's the only life I've known.
I'm just one man, living off the land.
It's more than what I do;
It's who I am. (Dietrich 1992, 46)

Emphasizing this self-conscious "pride" draws attention to the dynamic relationship between stigmatization and the politics of recognition. The swelling of movements that celebrate cultural pluralism or social diversity have found much of their inspiration in the experiences of historically stigmatized persons (people of colour, gays and lesbians, persons with disabilities, etc.). Taylor (1992) locates the demand for political recognition in the democratic right of all humans to dignity, which itself is related to identity.[6] His thesis is as follows: "Our identity is partly shaped by recognition or its absence, often by the misrecognition of others, and so a person or group of people can suffer real damage, real distortion, if the people or society around them mirror back to them a confining or demeaning or contemptible picture of themselves" (25). Moreover, Taylor argues that recognition is not simply granted to a minority or cultural group; it must be earned. It is won through social "exchange," where "we define our identity ... in dialogue with, sometimes in struggle against" others (33).[7] Note, especially, that (1) misrecognition has come to be associated with harm, and that (2) we define ourselves and our groups in dialogic association with others.

While stigmatization can have painful and alienating consequences at the personal level it can foster group self-definition for the purpose of constructing and pursuing political access. So when, for instance, Lorna or Beverly (the SOC members cited above) characterized themselves and their community as victimized, their statements operated not only as a claim to persecution, but also as acts of social self-definition, as invocations of kinship with like-minded individuals. Theirs was a call for group formation for the purposes of political recognition and, therefore, for power.

The speech event that fully exposes this dynamic intersection of stigmatization and political recognition is the controversial appearance of gyppo logger Mark Evans at the forest activist conference described in Chapter 3. Evans opened his presentation by invoking Oregon's highly controversial anti-gay Ballot Measure 9. As will be remembered, Mark endorsed gay rights. His subsequent comments, recorded here, represented him as unable to support the ballot: "There I was in the voting booth, and I couldn't do it. I know how it feels to walk into a bank and be laughed at ... to be a despised member of society ... I couldn't vote yes on Nine." This intimate appeal allied his own experiences as a stigmatized logger with the experience of being a member of the lesbian/gay community – a high-profile group that most environmentalists support. Mark utilized his stigmatization by employing it discursively to promote his group's worthiness of a similar quality and quantity of support.

The experiences of timber workers raise two central points. First, loggers' sense of having been stigmatized was indeed pervasive. Second, an understanding of what Taylor calls the "politics of recognition" allows for a contemporary reframing of Goffman, thus unifying the psychological burden of stigmatization with its socially constructive power. While timber workers saw themselves as victims, they did not, as environmentalists claimed, appear to be either passive or suffering from psychological ill-health. On the contrary, some members of timber-dependent communities employed stigmatizing attributions in order to foster their political and sociocultural ends.

Ultimately, stigma was negotiated within a dynamic tension born of the need of both loggers and environmentalists to achieve political power while maintaining their respective images as grassroots underdogs. On the one hand, environmentalists wanted to keep their opponents in check without undermining their own position by appearing to be uncaring towards workers. Harsh criticism of the industry was acceptable, harsh criticism of workers was not. Characterizing workers as pawns and members of pathological communities was useful because it implied concern while undermining the loggers' call for group and political status. Loggers, on the other hand, suffered from their association with big industry. Their grassroots movement was relatively new; their status as independent activists was in the throes of consolidation and was not yet secure in the public eye. Loggers did feel genuinely misunderstood, misrepresented, and dismissed, both by environmentalists and by the public. However, they attempted to use these stigmatizing experiences to cultivate sympathy, to reinforce their grassroots identity, and to gain political legitimacy through the construction of a logging identity worthy of respect and in need of protection.

Logger reading graffiti on a felled log. The graffiti reads: "Pardon me thou meek and bleeding piece of earth – that I am gentle with these butchers." The quote, which is from Shakespeare, appeared on a popular anti-logging poster.

Sign left by activist in what is actually Mt. Hood National Forest. The sign is a play on Oregon Senator Mark Hatfield (R), often criticized by environmentalists for his support of timber appropriation bills in Congress.

Photographs by Elizabeth Feryl

Road closure signs erected before a cut to limit access to otherwise public forests.

Heated exchange between the Oregon Forest Community Coalition members and environmentalists in southern Oregon.

80

Loggers demonstrating during President Clinton's 1993 Forest Conference, in Portland, Oregon.

A diverse crowd of anti-logging activists protesting a pending cut.

5
Voodoo Science and Common Sense

When President Clinton expressed his intent to find a middle-ground solution to the contest over the Pacific Northwest's remaining temperate rain forests, he emphasized the need for a scientifically viable forest plan: "Our efforts must be, insofar as we are wise enough to know it, scientifically sound, ecologically credible, and legally responsible."[1] Due to its overriding importance with regard to the survival of the northern spotted owl, many activists came to strategically position themselves on the side of science. In this sense, loggers and environmentalists occupied a shared (or figured) world in which both parties appeared sensitive to the privilege granted scientific explanations.[2] The need for a scientifically defensible forest policy was, at least superficially, a point upon which disputing parties appeared to agree: the language of science – sustainable forestry, ecosystems, biodiversity, wildlife habitat, wildlife biology, forest regeneration – populated much discussion about the forest. Yet scientific good intentions seem only to have confused matters: a short trip behind the public stage reveals the depth of activists' disagreement over scientific definitions of the forest.

Anthropologists have long noted that disputes about science are cultural disputes and that, as such, they are amenable to study as competing bodies of collectively held meanings and beliefs. This cultural designation has coincided with a flurry of intellectual activity across the social sciences known, generally, as science and technology studies (STS) (Franklin 1995; Hess 1992; Jasanoff et al. 1995; Nelkin 1987, 1992). Much STS work has focused upon comparisons of expert versus non-expert perceptions of science (Kraus, Malmfors, and Slovic 1992; Wynne 1992), and studies of public understandings of science (Hess 1995; Irwin and Wynne 1996; Wynne 1995).

Critical analyses of the expert/non-expert distinction have been fundamental to the deconstruction of modern science as "objective, analytic, wise and rational" versus public knowledge as "subjective, often hypothetical, emotional, foolish and irrational" (Kunreuther and Slovic 1996; Wynne 1995). Notably, Nelkin's (1992, 1995) work has challenged the oft-heard complaint that non-experts are scientifically illiterate; she has revealed the spectrum of moral and political values that shape both public and specialist positions. Others have sought to undermine characterizations of "non-experts" (or "the public") as a single unified mass by calling for the study of scientific discourses as they are put into practice by members of different non-scientific communities (Hess 1992). This permits an examination of similar knowledge bases as they are "used and reproduced in different local populations to provide grounds for their thoughts and actions" (Barth 1995, 66).

Academic preoccupations with public "misunderstandings" of science are, in part, a product of social analyses of the many public disputes about nuclear power and waste where we see a well-defined partition between nuclear scientists and an array of strongly anti-nuclear community groups. Yearley (1995), however, reminds us that, in most ecological debates, the distinction between expert and non-expert is less clear. Many environmentalists employ scientific knowledge to criticize the spoilage of natural systems. "Green" debates, therefore, provide fertile ground for re-evaluating public understandings of science. Indeed, when, as noted in Chapter 2, biologists began to argue that old-growth forests in the Pacific Northwest were imperiled, environmentalists quickly moved from discourses about the value of wilderness to discussions of mycorrhizal fungi and their links to the biological health of the northern spotted owl.[3] This appropriation of science ensures that not all science arguments are public/professional ones, but it does not ensure that different non-expert groups will assimilate expert knowledge to the point where they no longer criticize science; rather, analyses that fall easily into pro-science/anti-science polarizations conceal both the diversity of perspectives and the "substructures of ambivalence" found there (Wynne 1995, 383). Neither loggers nor environmentalists can be said to be definitely pro- or anti-science; participants in both groups express ambivalence towards science. As was shown in Chapter 4, ambivalence is itself a product of identity negotiations, of activists strategically constructing political advantage in relation to science and its accompanying hold on what counts as valid knowledge. Their understandings of science are thus best seen as statements of social agency, as challenges to the social authority of science, rather than as evidence of scientific illiteracy per se (Michael 1996; Wynne 1995).

The purpose of this chapter is to explore support for science as expressed by loggers and environmentalists. It will ask: What, in the case of the forest dispute, do participants from two opposing grassroots groups appear to mean, think, and imagine when they use the term "science"? And how is it that references to science and forests are linked to social agency? It will explore the diversity of activist thinking about science and forests in order to demonstrate the danger of reducing public understandings of science to contesting lay and expert opinions; rather, understanding will be seen to involve clashes between loggers and environmentalists as non-expert groups, each of which has ambivalent relationships to scientific knowledge. Public knowledge thus involves a triangular model of social process, where understanding is negotiated diacritically among lay groups in response to expert knowledge; it is a dynamic within which competing ties to identity become opportunities for social critique.

By and large, neither loggers nor environmentalists trust "science," though both are fully cognizant of the academy's reconceptualization of forests as multifarious natural systems. The result is a resilient scientific controversy over definition of the forest. Timber advocates tend to view forests as multi-specied crops that can be planted, harvested, and replanted in indefinite cycles. ("Trees are a crop. It's hard for humans to view it as that because it's so long term.") This is pitted against a science, advocated by many environmentalists, that views forests as enchanting, intricate, and enigmatic ecosystems not yet fully enough understood to reproduce once clear-cut. ("These old ecosystems are very complex. There is a lot more going on in those ecosystems than we [yet] know.")[4] Environmentalists remain ambivalent towards science even though it often supports their positions. They prefer abstract science to applied science because the former elaborates upon the complexity of nature's mysteries, thus promoting a protective approach to the forest. Loggers remain ambivalent towards science, rejecting abstract scientific knowledge in favour of notions of their own "common sense." They fear the political power of biology, which threatens their economic interests and their identities as farmers of the forest, yet they are willing to adopt technology that is the result of applied science.

Science and Environmentalists

Science, more than any other phenomenon, has forwarded the environmentalist cause. Forest ecologists have argued vehemently for harvesting techniques that cultivate species diversity and that eliminate the fragmentation of ecosystems. Wildlife biologists, as was documented in Chapter 2, have argued for preserving large expanses of old growth to accommodate a plummeting owl population. Thus, many environmentalists have had to acknowledge their good fortune, their winning of the intellectual lottery.

Yet, when environmentalists speak of the role of science in the forest dispute, a traditional emphasis on gaining greater knowledge through science is conspicuously absent; appreciation rests, instead, on the capacity for science to actualize a set of moral prescriptions that, heretofore, has been excluded. The very qualities for which science has long been revered – objectivity, rationalism, absolute truth – are not generally represented in the environmentalist discourse on science. Science is simply, and for the time being, a viable trump card, a good tool. It has broadened horizons, rendering non-utilitarian considerations legitimate.

Science of Wonder and Complexity

This pragmatic acknowledgment of the political utility of science is meaningful for many activists only when accompanied by an emphasis on science's more artistic dimensions. I refer here to a model of science that conjures up images of dreamy geniuses conducting abstract explorations of the nature of the universe.

Andrew Simon is the southern Oregon activist who, in Chapter 4, spoke against romantic, cowboyesque characterizations of loggers. During an engaging interview he narrated his transition from east coast intellectual to carpenter, Oregon resident, and forest advocate. His poetic descriptions capture those positive qualities that several ancient forest activists applied to science:

> True science is a deep and beautiful thing and has a lot to do with the evolution of intelligence and cultural forms. It provides the devices with which we step into the future. And there's a lot of very exciting science going on. Across the board we're stepping away from the older, reductionist notions of science and reaching for holistic pictures ... I mean to read David Bohm when he says, "Okay, there's three things out there. There's matter, there's energy, and there's intention." Alright. And it's the way those three things combine that determine what we have ... And this ends up all looking just like what is said in every classic religious scripture ... I mean, they're all on the same trip. Tremendously exciting.

Simon's enthusiasm is contagious, particularly with regard to his ability to conjure up an intangible, yet magical, aesthetic: the creative side of science. His comparison of science to religious texts and poetry denotes an artistic and philosophical love of science that elevates nature to a revered, untouchable (i.e., sacred) status. Similarly, the attention he grants science at its most abstract (i.e., the invocation of theoretical physics) anticipates a distrust of abstraction's opposite – applied sciences. Finally, Simon's sketch implies the need for science to accommodate two often irreconcilable standards: simplicity and complexity. He is extremely taken, as, admittedly, am I, by the structural and graceful simplicity of Bohm's idea that all life might come down to three basic elements: matter, energy, and intention. But he also covets a science that is holistic and/or anti-reductionist, which is to say a science that accommodates the infinitely complex. His is a science that agrees with the critiques of Newtonian (mechanistic) models of nature, that insists that very little about natural phenomena can be reduced to certitudes, or laws, that enable us to predict the future (Prigogine 1997).

Science and Mystery

Models of complexity, for some environmentalists, include a respect for the perceived ability of science to uncover nature's greatest mysteries – mysteries largely invisible to the naked eye. This enigmatic notion of complexity was initially difficult to grasp. It was not immediately apparent to me why Greg Norman, a builder and member of the Ancient Forest Grassroots Alliance (AFGA), dismissed forestry as "non-scientific" only to embark on a seemingly unrelated oration about recent discoveries in geology. I had asked Norman about his rejection of forestry as unscientific: Was it not simply, I queried, a matter of certain ideas in forestry becoming dated? This was not, for Norman, the pertinent point. He was more interested in forests (and science) as a metaphor for experientially unknowable phenomena. Hard rock geology provided an ideal example:

> I'm comforted by looking at major scientific transformations ... from old-school geology to plate tectonics. The idea that you could [go] out to the coast range in Oregon, say, and explain that geology in terms of what you see there [is] out the window. You can't do that. It's only explicable in terms of planetary movements ... Now I understand why. It [old-school geology] failed to explain what can be observed ... The kinds of global investigatory techniques that satellites and other things made possible were responsible, along with theoretical shifts, in allowing the larger picture to be seen.

Norman has faith in a science that can capture a holistic, or "global," perspective. But the majority of his ardour is directed towards science's capacity to comprehend the invisible. Plate tectonics provide an ideal expression of that which can be theorized yet cannot be discerned by the observer working in coastal soils or rock formations. Here, nature is beyond the reach of, or greater than, the powers of the human eye.

Michael Costas's narration of his conversion from logger to environmental activist also includes an abiding respect for the invisible. Michael talked of "discovering" ecology, a discovery that dismantled his previously held assumption that old growth was decadent wood, past its prime. This led to an appreciation of "species ... not readily apparent." Costas, a former logger, was sympathetic towards both the world of loggers and the world of environmentalists, but his frustration was explicit when he spoke of the capacity for science to reach beyond immediate sensory experience. "There is an arrogance on the part of rural people that because they live on the land they understand it ... [However,] without taking time to do research and look beyond just what our senses give us, we're not seeing the true picture." Truth, for both Norman and Costas, lies not in what can be seen but in what cannot be seen. Respect for science is linked to an interpretation of the natural world that is enigmatic, abstract, and experientially inaccessible.

Ambivalence and the Risks of Shifting Paradigms
A faith in non-sensory and abstract variables (which is characteristic of some

science), as well as a faith in aesthetic and political variables, is the basis for some activists' support of science. This conception of science is consistent with seeing the natural world as "complex." What is surprising is that, despite the kinds of scientific support I have outlined, and despite the fact that some of the most socially powerful voices in support of environmental protection come from the scientific community, environmentalists are fundamentally ambivalent about science. Science is seen as a fickle ally, at one and the same time an asset and a dangerous liability; this is because scientific support could shift at any time (Yearley 1993).[5]

This concern resonates in the thoughts of Julie Dawson, AFGA activist and editor of an affiliated environmental group's quarterly newsletter. "It's kind of ironic because of this whole thing with science. Now the environmentalists really have science on their side, whereas in the past it was actually the opposite." "The past" speaks to a history during which forest sciences served to mechanize and legitimize extensive logging (Hirt 1994). Old-growth forests were once dismissed as ageing, rotting forests in need of harvesting. "Sustainable" (often massive) yields from old-growth forests were justified on the grounds of erroneous expectations pertaining to the future yields of second-growth forests. Fear of the past, which is to say a fear that activist support of science might backfire, also recognizes the processual nature of scientific knowledge. Biological and ecological findings currently reinforce the complex web of life that sustains species, a paradigm that echoes environmentalists' (and Aldo Leopold's) call to protect the component parts of natural systems.[6] But what if peer review and continued empirical work produce less desirable scientific knowledge (at least from an environmentalist's point of view)? Recent work on biodiversity, for instance, suggests that a focus on particular species ignores dynamic systemic functions. It may come to pass that the extinction of certain species is systemically "normal" (Takacs 1996). If this were so, then would an argument (legally and scientifically) based upon individual species protection still hold? Changing arguments about biodiversity could become a new point of departure, a severing of environmentalist-science alliances.

Science, Ambivalence, and Emotion

When reporting on Forest Service whistle-blowers – employees calling attention to unlawful agency practices – Dawson also noted the similarity between journalistic demands for objectivity and the rational requisites of science. She expressed frustration with having to abandon the more emotional testimonials of whistle-blowers because expressions of this kind would be dismissed by policy makers as unprofessional or illegitimate. Here ambivalence was manifest in her belief that, ultimately, emotion, not science or reason, achieves the desired political effect. She noted that this is what propelled the ancient forest debate into the public spotlight. "This sense of credibility means you have to be factual and unimpassioned about what you're doing." Yet, "the people that made this a national issue are the radicals, the people who did use the emotional appeal of it." Rational courses of action pose, for Dawson, a challenge to her sense of herself as an activist. Accommodating science becomes the cognitive equivalent of "putting on a suit" to present one's case to persons in positions of power.

Some environmentalists also invoked "passion" and "limited vision" when questioning scientist-Green alliances and the embedded positions of human dominance found there. These invocations signal activists' trepidation concerning the balance between, on the one hand, the social authority science has lent to the environmental movement, and, on the other hand, environmentalists' "dis-ease" with technological society and the arrogant power implied by scientific ideation as an expression of mind over nature. The concern that science, as a knowledge (and rhetorical) system, might have too much power, along with the belief that the scientific view is intrinsically reductive, pervades environmental activists' uneasy embrace of science.

David Burns, Julie Dawson's activist colleague, offers a case in point. He responded to my inquiries about the role of science in the forest dispute with the following: "All of a sudden we're [environmentalists] defending scientists ... We can use it as a lever on the bureaucrats to get them to tow the line." Burns revealed a ready willingness to utilize scientific arguments that worked to environmentalists' advantage, but his support for science remained pointedly conditional. "It's good for now ... but it's a real limited worldview." Science is something to be tolerated, but it might, equally, become an unwieldy monolith capable of dominating the interpretative decision-making processes in the forest dispute. "Science is too big, it's too powerful ... we need to listen to other voices."

Humility and the Ecocentred Self

Appeals to polyvocal approaches to forests parallel anthropologists' attempts to redistribute the locus of intellectual power both socially and methodologically in the writing of ethnographies (Clifford and Marcus 1986). Similarly, environmentalists are attempting to reposition themselves in the human-nature exchange. In this renewed exchange, humility is the posture most often adopted in an attempt to foster an ecocentred self (Naess 1989) (I examine this in greater detail in Chapter 7). For now, it is necessary only to recognize that humility involves an attraction to the idea or the hope that there is something in the world of nature that is bigger than oneself. Humbly juxtaposing self and nature helps to deconstruct tenacious nineteenth- and twentieth-century notions of "man [sic] dominating nature" and suggests a relationship between humans and nature wherein the latter is perceived as infinitely wiser and more powerful than the former. Only human arrogance (and, by association, scientific knowledge) has assumed anything else. Again, in the words of activist David Burns,

> Nature will recover and that may take us all out ... I saw this thing on comets. It was actually a videotape of a comet streaking through the upper atmosphere. If that had hit the Earth, it would be the end of life on Earth. For higher life forms. Somehow I just felt so great about that ... It's like, we can screw up all we want and all we're really going to do is screw up ourselves. That's great, because nature will always overcome; I can just relax a little bit with knowing that.

Expressions of humility hold a vague emotional charge and hint at an affective state that may be glossed as enchantment – as being under a spell, as experiencing a feeling of great charm or fascination.[7] It might be equally appropriate to think of environmentalists as fearing disenchantment, a state that writer Robert Michael Pyle (1996) describes as involving the loss of the ability to imagine the fantastic. To lose the land, writes Pyle, is to lose the capacity for imagination. When attempting to make sense of the fantastic with regard to speculations about Bigfoot, Pyle refers to scientific hubris and offers the following caution: "When the topography is finally tamed outright, no one will anymore imagine that giants are abroad in the land" (17). Also apparent in Burns's comments above is a degree of relief and satisfaction in the idea that nature will have its revenge. A strong undercurrent of guilt accompanies the attractiveness of the extinction (at least theoretical) of one's own species. In the court of nature, humans have exhausted their impunity.

The intensity of Burns's desire to be humble towards nature results in his ambivalence towards science. The plea for other voices and the fear of science-dominated courses of action must be understood in relation to this appeal for human humility towards nature. Fairly or not, within this broad cluster of ideas, science is associated with dominance over nature and the possible loss of enchantment. Historically, science's objectification of nature has jeopardized the hope of adopting a more appropriate human posture towards it. In other words, it appears that a preoccupation with science potentially counteracts relationships with nature based more fully upon humility and emotionality.

Voodoo Science

Finally, distrust of science cannot be separated from distrust of forest management. Burns's search for humility implicitly seeks permission to find solace in the knowledge that nature – unmanaged – will be allowed simply to exist. He thus pairs his emphasis on passion with a reference to the management of nature, where management is synonymous with clear-cutting. Cynicism towards management is common to the degree that there has evolved a widespread ironic use of the term, typically expressed by a hand gesture indicating quotation marks.

> It's a passionate thing. You know, passion for nature. There's a lot of logging going on now in the Tongass National Forest, and I've actually never been [there] ... I just know that I'll never see it like it was, and I just get really upset about that. Because there's a real pristine area. Now, there's gonna be clear-cuts; you know, it's going to be "managed." I just need to know there's wild land.

Critiques of management are also critiques of modern science, and they derive from the legacy of Gifford Pinchot. Pinchot, first chief of the Forest Service, was the German-trained forester whose views were outlined in Chapter 2. As a populist with close ties to Theodore Roosevelt, Pinchot promoted the accumulation of national forestlands, the American public's natural resource trust. Yet, as a forester, he ushered in an era of scientific management wherein forests were seen as sites of production, engineered for maximum efficiency and minimum waste (Hays 1958; Hirt 1994). The subsequent cutting and rationalization of national forestlands, and the science that legitimized the movement, is the basis for most science-management skepticism among environmentalists. "Forest management" has become a euphemism for embittered discontent with the literal and metaphorical "machines in the garden," and the clear-cutting they enable.[8]

Disciplines that have moved away from modern science have not necessarily eased the distrust of environmentalists. Ecology is a postmodern, or at least late-modern, discipline. It rejects totalitarian control of nature in favour of an emphasis on chaotic and non-linear relationships among the component parts of natural systems (Downey and Rogers 1995; Yearley 1996). Nonetheless, many environmentalists view "applied" ecology as so much jargon – an insubstantial shift in language from talk of forest-crops to talk of ecosystemic management.

Andrew Simon's earlier cited lyrical monologue on the beauty of science was accompanied by utter disdain for ecosystem management.

> Unfortunately right now as I look at the forest issue, we're talking largely voodoo science. We're talking arrogantly, as if we know about things that we've never done. No one has ever long-term managed an ecosystem, but we've got professors of ecosystem management ... Under these circumstances one should be *extremely* conservative in being part of any disruption.

Simon recognizes his ambivalence towards science, its horror- and awe-inspiring products: "I'm both excited about [science] and scared to death of it." But when having to make a choice about actualizing the creative knowledge he so admires, his values are clear. He will not support any application of "voodoo science"; that is, of applied or forest-managerial science. Within the climate of the Pacific Northwest timber wars, "extremely conservative" generally means: No cutting, no human disturbance of the forests in their current form.

To summarize environmentalists' ambivalent response to science, I shall defer to Elger's oft-cited assertion that "ecosystems are not only more complex than we think, they are more complex than we can think" (Dietrich 1992, 110). It is an appropriate aphorism to recall when considering environmentalists' affinity for the intricacy of nature. If nature is more complex than we think, then it cannot be dominated by the human mind. This definition of science implies a dismissal of applied sciences, provides further opportunity to amplify issues of humility and enchantment, and signifies the subservice of humans to natural forces. A science that is desirable is a science that is, like nature, idealized, beautiful, mysterious, and complex – but not dominating. Therein lies the desire of environmentalists to deconstruct arrogance towards nature and their concomitant preference for an ecologically focused identity that cultivates humility.

Science and Loggers

Timber Advocacy, Science, and Common Sense
Having gone into considerable detail as to just what environmentalists think and say about science, I must add that a proportionate sampling of timber workers' views was not readily available. My indirect, and finally direct requests to timber workers for commentary on science were invariably met with defeat and disgust – "I don't want to talk about it," "such idiocy," "I'm so sick of science." Most environmentalists' conceptions of science are anathema to timber workers and their advocates. Science, for most loggers, is not a set of ideas that one looks at, explores, and turns over in one's mind for the benefit of the tape-recording or note-taking anthropologist. Science, as imagined by most loggers, is valued if and only if it is also experientially driven and, therefore, practical, accessible, visible, and/or tangible. In the world of loggers, science must make common sense.[9]

Often juxtaposed with science, common sense remains a slippery though ubiquitous construct associated with social and intellectual movements as diverse as freedom from colonial England (for eighteenth-century republican Thomas Paine) and pragmatism (for philosopher William James). Einstein, meanwhile, defined common sense as both the refinement of everyday thinking and the collection of prejudices acquired by the age of eighteen. Common sense is also the quality that neo-Marxist scholars ascribe to hegemonic discourses – particularly those that frame traditions, morality, codes of behaviour, and so on as conventional wisdoms against which alternative views appear odd, subversive, or simply incorrect (Gramsci 1971). A quick Internet search reveals the spectrum of political organizations that promote the cause of common sense, some decidedly right wing, others less so.[10] Among the better definitions of common sense, however, remains that offered by Clifford Geertz (1983). He outlines several features (what he calls "quasi-qualities") of common sense. Some definitional overlap can be found across this list, but the composition of three of his dimensions accurately encompasses the majority of timber workers' assessments of science. They are "thinness," "accessibleness," and "practicalness."

To understand this assertion, consider the following two patterns of response that I invariably encountered when discussing science (and its role in the forest debate) with timber advocates. The first pattern, which comes under "accessibleness," maintains that, in order for scientific knowledge to be acceptable, its knowledge claims must also be evident ("seeing is believing," to put it colloquially). The second pattern, which comes under "thinness," speaks to the assumption that there is a self-evident reality to all matters: "Everything is what it is and not another thing." That which matters, that which one needs to know, is readily apparent, not hidden: "The really important facts of life lie scattered openly along its surface, not cunningly secreted in its depths" (89).

In the first pattern, the logger-speaker typically tells of his or her faith in the capacity of harvested forests to reproduce themselves by telling the story of a naive outsider (typically urbanite, non-Pacific Northwesterner, or easterner) who "arrived, saw [for him- or herself], and believed." It is a response generally meant to contradict what loggers glossed as "doomsday science."

Beverly Mason and her husband Vern own a small logging company that employs a crew of six men. As noted in Chapter 4, they are active members of Caledon, Oregon's, Save Our Community (SOC). When discussing science, Beverly recounted for me an observation made during a drive in rural Oregon. She and Vern had come upon a maturing stand of trees in a nearby national forest Vern had logged about thirty years before. "One time Vern and I were going for a ride, and I had never been back near Brookside. I thought going up and down I-5[11] was Oregon. And I couldn't believe the trees out there, I was just dumbfounded that there were that many trees in the world." In this setting Beverly casts herself as a shocked outsider taken aback by the abundance of trees available to the naked eye.[12] Adjacent to Beverly's "shock" at the plethora of trees is her reference to I-5 drivers (i.e., probably urban and, by implication, environmentalist) who never bother to explore rural Oregon, never bother to see the forests for themselves. There is a firm belief in timber communities that a tour of the back country would, by sheer force of visual evidence, convert any outsider. In the language of common sense: "Everything is what it seems to be." Clear-cuts grow back. Evidencing their commitment to forestry as farming, and to trees as crops, members of timber communities do not speak of trees versus ecosystems. "Forest" means second growth and old growth, and, as far as most timber advocates are concerned, it's just a matter of time before second growth becomes old growth.

Typically, timber supporters see second-growth forests as more aesthetically attractive (with their straight lines and even stands) than old-growth forests. The second growth in the stand logged by Vern is uneven and included one particularly robust patch of trees located on the old logging road. Vern explained that the taller, more successful second growth is taking place where the skid road used to be. Skid roads are the short logging roads that bear the weight of heavy machinery transported to, and used at, individual logging sites. Science-based forest policy imposes restrictions designed to eliminate the ground compaction produced by road building, log dragging, and heavy machinery. A great deal of time and expense are taken to ensure that trees are removed with as little ground depression or gouging as possible. Beverly described what Vern had said about the previously logged area. "Well that skid road he was showing me, he said, 'Where you see the taller trees, that's the best growth, where the skid roads were.' And he said, 'They're telling me these days, you know, keep off the ground because it'll hurt the tree growth.' I thought, now there's a stupid logger, you know – practical!"

Beverly's story expresses her frustration with a science that makes little experiential sense to her or to her husband. Pointing to the taller trees is a way of saying: This is all the evidence I need; science needs to corroborate my experience. Observation and experience tells me that soil compaction helps rather than hinders growth. Beverly develops this point with a sarcastic allusion to the "stupid and practical logger," calling attention to the subservience loggers feel they are expected to adopt towards the Forest Service's seemingly nonsensical scientific guidelines.

Beverly's story elevates the value of visual observation and practical experience while simultaneously downplaying scientific expertise. This is the essence of the accessibility dimension in Geertz's definition of common sense. A basic feature of common sense is its accessibility to everyone. Its tone is anti-expert, if not anti-intellectual, thereby rejecting explicit claims to power: "There is only what we rather redundantly call experience" (Geertz 1983, 91).

Vehement adherence to a common-sense framing of science, including the "thin" idea that "seeing is believing," is evident in Lorna's model of appropriate knowledge. Lorna, again, works for Mason Logging and is an active member of SOC. We had been talking about the various panels of government scientists that had been convened to study the plight of the northern spotted owl. Lorna's references to expert knowledge involved an explanation of events wherein a photographer becomes, by virtue of his photographic eye, a scientist. Specifically, Lorna had been criticizing the absence, on the federal government's scientific assessment team, of biologists who could substantiate both a healthy spotted owl population and the presence of owls in second-growth forests.

> That isn't right. If you've got a biologist over here that's saying "There just aren't any [owls]," then if there's a biologist over here that says, like the photographer from Bend, he said "I have spent the last ten years living in the forest, I take pictures of these, this is what I do for a living, the spotted owls are everywhere" … But do you ever see that side? These people are never allowed to put their foot in the door.

Lorna's belief in empirical, photographic evidence is intrinsic to her frustration with scientific experts. A photographer is, therefore, the perfect stand-in for a biologist. Lorna's rules for adequate knowledge dictate the comparability of these two roles. The photographer and the biologist are put forth as equally valid, if not indistinguishable, custodians of knowledge. Expert science, as opposed to common-sense science, is identified as a source of disenfranchisement and loss of control over the political processes that govern one's life; the common-sense scientist, symbolized here by the photographer, is "never allowed to put [his/her] foot in the door."

A dismissal of science as inaccessible is central to loggers' critiques of science. The dissent of loggers incorporates a specific perception of science as a closed domain – an elite intellectual (and political) body bent on imposing abstract models of operation and protection on forests that scientists, so the accusation goes, have never seen. When the Clinton administration sequestered approximately 100 scientists, economists, and social scientists in a Portland hotel to produce a draft forest plan, members of timber-dependent communities reeled with resentment. This was precisely the point loggers felt they had been making for months. Life in timber communities had turned into an endless parade of ivory-tower experts with flow charts and visual models aimed at reducing the timber supply they had come to depend upon.

For timber workers, resentment escalated in the absence of opportunities to test and to apply new plans. It seemed to them that people without any experience in the woods were exacting an absurd and fatuous set of abstractions. Clinton's sequestered scientists simply represented the largest gathering of experts to date, an insult on an even grander scale. Consequently, among loggers the team of scientists came to be known as "the tower of power," "the cartel," and "the great fortress." Accompanying these bitterly sarcastic labels are images of rebellious vassals (i.e., loggers) chafing under the constraints of academic dominance.

Timber Advocacy and Praxis

Martha Grant has put her considerable community-organizing skills into running a grassroots timber group known as the Silver Valley Timber Action Coalition, a subset of the Oregon Forest Community Coalition. Her husband drives a logging truck. During one part of a conversation held in her cramped office, I asked, without referring to science, what her ideal solution to the forest dispute would involve. She replied by drawing a portrait of science as being burdened by intellectual inbreeding, a product, she argued, of a dearth of applied experience.

> The first thing I would do is I'd get rid of all these government scientists. What we've done – what the government's done – time after time is stick the same scientists in the same room and say, "Well, the other plan didn't work, come up with a new one." And one of these days, you know, at some point, they're going to realize that these same scientists keep coming up with plans that don't work. So we need the new, the innovative, those that have actually been out in the field and implemented.

Technologies based in experimental implementation, knowledge that can be applied, are what Grant values. The implied antithesis of her call for implementation is embodied by the numerous plans and abstract constructions lying unused or unusable on the desks of government knowledge brokers.

Understanding timber advocates' conception of science is easiest when thinking in terms of praxis (Bourdieu 1977), habitus (Bourdieu 1985, 1990), and common sense (Geertz 1983). Praxis comes from a Greek word meaning "doing," or "action," and privileges practical action in the discovery of knowledge. Habitus is that knowledge acquired through labour – knowledge that, in time, becomes "second nature." With regard to loggers, how one labours both imparts and embodies the full history of working the land. Such knowledge, following historian Richard White (1996, 179),

> is connected with physical experience, but it is not derived solely or often even directly from physical experience. Working communicates a history of past work; this history is turned into a bodily practice until it seems but second nature. This habitus, this bodily knowledge, is unconsciously observed, imitated, adopted, and passed on in a given community. Our work in nature both reinforces and modifies it.

Experientially driven knowledge – praxis becoming habitus – rather than scientific knowledge, expresses class and cultural identities unique to loggers.[13] The invocation of such knowledge promotes latent class values in terms of hands-on, or manual, labour and simultaneously reifies loggers' identities as farmers of the forest.[14]

The ready tendency for timber workers to dismiss abstract knowledge was directly confronted during a conversation I had with Steve Fuller at an old-growth mill outside Eugene. Fuller, a logger, defines the relationship between forest management and treatises on forestry as antagonistic. His valuation of his own skills reduces the importance of academic knowledge, extolling instead the education provided by a childhood in the forest and a lifetime of working there.

> You know, I don't know – ever since I was a little kid I've known how nature works. And I know, [from] working in the woods, how the forest works. I haven't learned that from reading about it or modelling it. You have to experience it. You really do. And the so-called experts, they don't know any different. When they come up with theories or reasons for things, they don't know if it's right or wrong because they haven't experienced it. I can look at it and say that's a bunch of crap, that's not how it works. And I do that a lot.

Steve's opening use of "I don't know" (as well as the shrug that accompanied it) is typical of the self-effacing manner I frequently encountered in timber communities and at logging sites. Intellectual and/or social arrogance is a cardinal sin among loggers. Despite this manner, Fuller is insistent about his confidence in an experientially acquired knowledge base. The consistency with which knowledge that can be labelled abstract, or at least academic, is dismissed was consistent with my own experience of gaining credibility from timber workers as compared to environmentalists. Among loggers my credibility as a researcher was dependent upon (1) being a quiet listener/observer, (2) my physical agility (e.g., the quickness with which I was able to move up and down hillsides), and (3) my willingness to accompany a logging crew or visit mills in Oregon. Comparatively, I gained respect from environmentalists by the cleverness or uniqueness of my questions and by my ability to provide an engaging or mildly challenging intellectual, and sometimes emotional, interaction. This was especially true when interviews took a subject to some mental domain not yet "thought of" or explored.

Doug Newman, Steve's co-worker, took me to several cutting sites. Gesturing out the window of his truck, Doug proposed that their boss, a quiet, wiry, greying man in his mid-sixties with fifty years of experience as a lumberman, settle the timber dispute. His argument: "He listens, he knows all its corners. [Doug was pointing outward toward the Willamette National Forest in which we were driving.]" To "know" is to do. Doug's comments were meant to reiterate his belief in the logic and wisdom of experience (i.e., knowledge) that has or can be applied. It supersedes insights generated from any other source. The distinctly anti-intellectual tone is also an attempt to equalize and/or prioritize skills infrequently valued by the academy.

The widespread likelihood of timber workers and advocates invoking images of experience-based skills is also a way of defending the social identities that accompany those skills. I have already mentioned the propensity for timber workers to define themselves as farmers of the forest. Hard work and knowledge born of continual episodes of trial and error in particular forests are fundamental to timber workers' self-assessments regarding who they are and what they are good at (James-Duguid 1996). Thus Brian Wynne (1992) found the judgments of risk and of science on the part of English highland sheep farmers to be intimately tied to their social roles. This explains why, after a lengthy discussion of science, Steve Fuller attempted to explain the behaviour of environmentalists in terms of an attack on his skills, his productivity, his work, and his identity. His explanation includes a surprising contention that environmentalists are jealous of loggers and clearly favours expertise produced and sustained by experience. I close this section with Steve's intriguing comments at the end of my interview with him. Notice, especially, how he implies that environmentalists, not loggers, suffer from a kind of alienation. Environmentalists are portrayed as being isolated from the products of their labour. In sum, Fuller has managed to elevate the value of knowledge based in practical action while resisting the usual Marxist pairing of alienation and manual labour.

I philosophize a lot. I mean I think there's people, there's people like ourselves that produce. We're independent, we can work hard with our hands, with our minds, and produce something physical. There's other people that are inadequate and they're insecure in their inadequacy. And I really believe they're jealous of the producing society, the independent society that can survive on their own.

The world could end tomorrow, and as long as it wasn't burned up, I could live. The preservationists couldn't. I could live, I could eat. I could shelter myself, and my family. And I think there's a lot of jealousy on their part. Now I don't know if jealousy is the right word.

[But] they know we work very hard. We may not be as educated as they are. But we can sustain ourselves at any point in time. They can't. And I think they resent that. So I think there's an intent on their part of "I don't like these people." They may be [thinking] in the back [of their heads], subconsciously, [that loggers are] better at surviving in the game, you know, in the game of life. "[So,] let's make it so they [loggers], you know, they're not doing that [logging] anymore."

I'm – you know, that's just something that rattles around in my brain. Now I don't know if that makes any sense or not. And it's subconscious I believe on their part. Yeah. I'm not alone in saying, you know, I could survive. Because everybody in this industry is very self-sustaining, very independent, have learned to do what they need to do to live. And you think about that and, you know, think about that and see if you see that.

And ... the preservationists, what do they do for a living? They're not necessarily happy with what they do because they can't, like I said, they can't see what they've produced in that day or that week or that month, [but the desire] it's still in them. It's still in them. They need to be able to do that. It's just something they resent, the fact that we can.

I always try to understand why people do what they do. And, you know, sometimes I think I'm right and sometimes wrong, but I think that's a point.

↶

Postmodern scholars have argued effectively that "knowledge is never simply knowledge of something, it must also be knowledge for someone" or, I would add, some end (Downey and Rogers 1995, 273). Environmentalists might be more likely to endorse science because of its acknowledgment of the mystery and sanctity of the natural world; and certainly there is evidence of their in-depth knowledge of some science. This may strike those wringing their hands over the public's failure to understand and take up science as a victory, but the "victory" is shallow. What is most revealing is the ambivalence of environmentalists towards science. Recurrent ambivalence emphasizes their uncertainty about where, exactly, to place an eco-centred self in the field of uncomfortably authoritative, yet potentially advantageous, knowledge. An abstract, deeply ambivalent, and anti-applied image of science is entirely consistent with a belief in the need for humans to maintain a humble, unintrusive stance towards nature. Consequently, for environmentalists, endorsements of science vary, being part of the larger process of reconciling contradictions between humility and the power-over-nature affiliated with science.

Loggers, too, are caught up in a political process that compels them to speak through an idiom they might otherwise choose to ignore. Given the collapse of modern science and its contention that forestry should be conducted as a rationalized agricultural process, loggers are left to negotiate their own identity-based critique of expert knowledge. A critical conception of science as non-commonsensical, even to the point of standing in contradiction to experience, and publicly inaccessible protects timber workers' identities as utilitarian producers, as farmers of the forest.

Ultimately, both environmentalists and loggers remain ambivalent towards science and work to promote distinct, competing challenges to its authoritative grip. Activists need to attend to the charismatic force that is science. Ironically, that attention may become the basis for science's undoing because activists operating in the public domain must address science, and this may, in turn, lead to its democratization (Weeks 1995; see also Beck 1992). In the case of the forest dispute, references to science become vehicles for a critical lay dialogue about knowledge and philosophies of conduct towards the natural world. Social studies of environmental conflict and of science might benefit from a parallel theoretical reorientation. A shift from a lay-expert dichotomy to a triangular model of social process is warranted, for this could take into account lay rivalries about scientific discourses and how they pertain to nature. These rivalries embody important public efforts to diacritically rewrite the criteria for determining valid knowledge; consequently, they should become central to analyses of public knowledge as it pertains to environmental debates.

6
Theorizing Culture: Defining the Past and Imagining the Possible

The last two chapters focused explicitly on the broad undercurrents of thought and action that comprise activist efforts to achieve and maintain grassroots status and dispute dominant scientific definitions of the forest. Their identity-driven agency is conveyed to the extent that both parties contest the situational applicability of entrenched cultural institutions; namely, democracy and science. But the point for all activist groups is not simply contestation of, or resistance aimed at, dominant cultural forms. It is, equally, their creative, imaginative, and often idiosyncratic capacities as cultural producers, their ability to suggest new alternatives that may or may not become the basis for new cultural forms. It is to the synthesis of contesting-resistant and imaginative-productive practices that the next two chapters turn.

The ancient forest conference described in Chapter 3 opened with a prayer delivered by a Takielma woman and a Karuk man from northern California. Both invoked "Mother Earth" and the accompanying need to "speak for the voiceless inhabitants of the forest." The presence at major environmental conferences of tribal people has become, in Beth Conklin's phrase, "almost de rigueur." Amerindian voices "are now heard, and faces are now seen, to an unprecedented extent" (Conklin 1997, 721). In some cases, participation is confined to a conference's key ceremonial moments, while in others Aboriginal and non-Aboriginal collaboration is more widespread in that the conference arena operates as a hybrid "middle ground" territory wherein the goals of both parties converge for their mutual benefit.[1] But conference events aside, the forest community and ancient forest movements studied here were comparatively devoid of either a local or a non-local Native American presence. The want of Aboriginal participation was especially striking alongside the propensity of non-Aboriginal activists to speak frequently and in some detail of Aboriginal practices.

It is a discrepancy that deserves attention and, thus, constitutes the explicit focus of this chapter. I will begin with an examination of the "noble savage" legacy – a dominant discourse about the "nature" of culture – that underpins the tendency for White activists (among others) to valorize Aboriginal peoples as ecologically righteous. Thereafter, activists' Aboriginal-centric references will be examined in order to ask: What, in fact, is at stake in the logger-environmentalist dialogue about past peoples? How is the dominant (and essentialist) narrative in Western culture concerning the ecological nobility of Aboriginal peoples situationally transformed to meet the needs of non-Aboriginal activists? Why is it that activists both disparage and elevate Aboriginal peoples? And what are the implications of this dialogue for Aboriginal peoples when they are neither a major force in the Oregon movements described in this book nor present to refute or defend White activists' portraits of "Indianness?"[2]

The subject of nobility, as expressed in primitivist theories, is an uncomfortable one for me. An anthropologist by training, I am only too aware of my discipline's promotion and demotion of the "nobility" thesis and its ecological variations. My first impulse was to avoid the topic, but this would have been ethnographically dishonest. It would have permitted me to hide from the activist predilection for playing off of and, at times, epitomizing White assumptions that Aboriginal peoples innately and effortlessly manifest a lifestyle and worldview that is closer to the conservationist ideal than any other (Buege 1996; Conklin 1997; Grande 1999; Waller 1996). It would also ignore the fact that nobility and anti-nobility discourses are a singularly important window through which to examine the centrality of cultural legitimacy to the forest dispute. As it turns out, activists' repeated mention of "past peoples" and "other cultures" speaks directly to their own ethno-theories about who they – Oregonians, or, more broadly, Americans – are, were, or may some day be. It is the dialogue through which activists assert their cultural authority and, thus, their right as a people to determine land-use practices, to inscribe human culture on the land.

The Noble Savage Legacy

From the Enlightenment forward, European philosophic and artistic traditions tended to juxtapose the perils of an increasingly complex civilized world with the benefits of "primitivism." It was common to employ the noble savage tradition to contest the absence of civil liberties and the persistence of hierarchical systems of power, and to use imagined and actual Aboriginal traditions as a ready basis for critiquing Western society's own record of social and ecological failure.[3] The nobility assumption, following Berkhofer (1978), is that there are simple people, living in an Edenic landscape and gentle climate, whose powers of reason and ability to live in harmony with nature ensure relief from the evils of civilization.

By the late eighteenth century, representations of nobility focused upon Aboriginal peoples' emotional, or "romantic," sensibilities. A preoccupation with Aboriginal peoples who were presumed to be exotic and close to nature had long been present, but the newer romantic Aboriginal was conceived as apprehending the world through feeling, emotionality, and an acute sensitivity to a lush and bounteous physical world. Jean-Jacques Rousseau, a figure popularly associated with these ideas, promoted the desirability of spontaneous expressions of feeling and, most important, the internalization of (today we might say the bodily living of) the human-nature relationship.[4] The Romantic scholar and artist's ambition was to write or paint life and nature as poetic and inspirational, his or her goal being to evoke in others the ability to "feel and imagine nature and life deeply" (Berkhofer 1998, 79).

Conservative social forces often contested portrayals of other peoples as noble because they recognized the polemical benefit of idealized portraits of non-European societies. Then, as now, those defending the status quo, including the supporters of church and state institutions who were the target of primitivist alternatives, frequently attack the noble savage premise by pointing out the "brutish existence led by contemporary primitives" and by strategically employing examples that counter "the natural goodness and happiness" of Aboriginal peoples (77).

Primitivism in Anthropology

Both before and throughout the twentieth century, the discipline of anthropology has done much to contribute to the maintenance of primitivist stereotypes. The ecological nobility of past and contemporary peoples remains a subject of heated debate and will likely remain so for the coming period. An exhaustive review is not warranted, but it should be noted that three distinct fields of scholarship – nineteenth-century social evolutionary theory, cultural pluralism from Benedict (1932) to Geertz (1973), and human ecology from Rappaport (1968) forward – all helped foster the idea of Aboriginal peoples as exotic, ecologically noble, and, in some cases, historic relics.

The notion of past peoples as relics of a prior period is most often associated with colonial-era scholarship and mid-nineteenth-century evolutionary theory. Social scientists of this period (e.g., E.B. Tyler, L.H. Morgan, J. Frazer, etc.) introduced multiple theories of societal development, many of which sought to fit "primitive" cultures into an evolutionary schema. Groups were hierarchically ordered according to their likeness to White Europeans; that is, they served as indicators of earlier, or "lesser," stages of cultural development en route to the cultural apex epitomized by European civilization. Portraits of cultural exotica (particularly as they concerned animism, totemism, and mythology) were conspicuous in the evolutionary literature, whereas the nobility theme, for the most part, remained dormant until the modern era of anthropology.

Recognized for undermining the racist assumptions of evolutionists, modern anthropologists were, in no small part, motivated by the opportunity to supplant evolutionary thinking with the tenets of cultural relativism. Cultural relativism maintains that the world is populated by many unique cultural groups and that no one group or cultural system is better or worse than another. Distinguishing between genuine (closest to the past and/or operating as self-contained wholes) and spurious cultures, and creating clear physical and cultural boundaries between Aboriginal cultures by isolating one's study to one particular group – the Nuer as distinct from the Dinka, the Nootka (Nuu-chah-nulth) as distinct from the Kwakiutl (Kwakwaka'wakw), and so on – was an important means of reifying the diversity of human culture.[5] A second means of doing this involved (and still does) the privileging of studies of culturally isolated and geographically distant people. Studying cultures that were remote and different from mainstream American and European peoples became central to the anthropologist's credibility.[6]

Most anthropologists of the post-1960s generation, myself included, continue to uphold the fundamental importance of respecting cultural relativism. Ruth Benedict's (1932) message concerning the cultural relativity of normalcy remains a compassionate plea for tolerance and a stellar example of 1920s and 1930s anthropology as social critique. Learning about others abroad was a means of suggesting how social life could be differently construed at home. This critical function remains central to the discipline (Marcus and Fischer 1986), but it also leaves anthropologists vulnerable to the promotion of some cultures as not only unique and suggestive of alternatives, but also as better, genuine, even noble.

In the work of human ecologists, subtle and not so subtle expressions of nobility took on ecologically specific meaning in the late 1960s and early 1970s (Rappaport 1968; Vayda and McKay 1975). Influenced by growing evidence of environmental degradation in industrialized nations and the need for critical examination of this, the cultural ecology field was founded on two assumptions: (1) that tribal customs, many of which seem illogical or bizarre to Westerners, function ecologically to keep local populations adapted to their environments and (2) that primitive peoples lived in near perfect equilibrium with their niche environments because culture acted as a brake against destructive human impulses that threatened ecological sustainability (Biersack 1999; Headland 1997). Industrial nations were rightly criticized by this subfield's proponents for operating beyond their ecological means; however, its conclusions further valorized small-scale, and especially primitive and Aboriginal, cultures as noble.

The questionable veracity of claims about niche adaptability and growing interest in culture-nature interactions helped spawn several more subfields (e.g., political ecology, historical ecology). More recently, scholars from anthropology and related fields in biology have treated nobility as a serious hypothesis in need of testing rather than as a popular discourse about small-scale societies. Some seek evidence, therefore, to dismiss the ecological nobility thesis as "the golden age that never was" (Diamond 1992). Redford (1991), significantly, coined the term "ecologically noble savage" in an article so named in the periodical *Cultural Survival Quarterly*. He and others (e.g., Alvard [1993]) went on to challenge the belief that contemporary Aboriginal peoples resist behaviours that are environmentally unsustainable, presenting, instead, evidence for ecologically dysfunctional practices.[7] "There is no cultural barrier," writes Redford (1991, 46), "to the Indians' adoption of means to 'improve' their lives (i.e., make them more like Western lives), even if the long-term sustainability of the resource base is threatened. These means can include the sale of timber and mining rights to Indigenous lands, commercial exploitation of flora and fauna, and invitations to tourists to observe 'traditional lifestyles.'"

The point, for Redford, is that, while some Aboriginal peoples have acquired a vast body of empirical knowledge about the physical world, they are not, as a category of people, automatically disinclined to adopt practices that ensure immediate physical survival or prosperity simply because such practices would threaten local resources. A prime example, according to Redford, is the near universal Amerindian adoption of firearms for hunting in the neotropics, but the principle applies equally to the adoption of multiple Western industrial resource extraction practices. Redford is, however, overstating his case when he denies the possibility that humans can act in both self-interested and nature-enchanted ways. The crux, for most anthropologists, is that they recognize as possible both case studies of ecological degradation brought about by Aboriginal populations and case studies of superb land management by Aboriginal peoples (Anderson 2000). For others, the point is not ecological nobility per se but, rather, knowledge as evidenced by the revitalization (but not valorization) of studies of traditional ecological knowledge (Biolsi and Zimmerman 1997; Dove 1996; Ellen and Fukui 1996; Goodenough 1996; Nader 1996; Sillitoe 1998a, 1998b). These efforts continue to produce examples not only of profound empirical sophistication among so-called primitive peoples, but also reveal possible solutions to some of our more vexing environmental problems.

In the end, the debate remains sticky (and ironic) due to the uneasy coexistence of two trains of thought: (1) the belief that anthropologists have been presumptuous about ecological nobility and (2) the finding that Aboriginal activists have wisely, strategically, and sometimes ironically exploited for their benefit Western assumptions about noble non-Western peoples. Anthropologists are increasingly self-conscious about their past assumptions regarding nobility, while, at the same time, they support the right of Aboriginal peoples to achieve cultural revitalization through the use of "nobility" discourses and/or the building of First Nations cultural centres that emphasize a "primitive" past (Brosius 1999a; Clifford 1997; Conklin 1997).[8]

Essentialism and Ecological Nobility

Ultimately, the features of the nobility discourses remain central to this book because they presage the criteria that forest activists use to defend their own depictions of an ecologically (and culturally) defensible future. The echo of nobility discourses affiliated with Enlightenment, Romantic, and anthropological traditions permeate activists' thinking, producing frequent speculations about past societies thought to have "lived in harmony with nature" and/or to have been innately attuned to and protective of the non-human world. There is an undeniable penchant among many environmental activists for imagining matriarchal societies worshipping the Goddess or agricultural societies whose impact on the land is believed to have been minimal: "We are asked to imagine long lost societies and to envision creating a twenty-first century in which ecological awareness is as important as it was then" (Buege 1996, 72). Support for multiculturalism in liberal American thought has generally resulted in updated (though not necessarily destereotyped) imaginings of most minority groups, whereas Native Americans remain trapped in the "time warp of American iconography" and are thus still crudely imagined as buckskin warriors and exotic maidens (Grande 1999).

Within the context of popular culture, we are still apt to hope that somewhere "out there" lives a people relatively untouched by time and technological advance (Headland 1997). We assume that such people possess extrasensory perception when reading, mapping, or comprehending the physical world. Marie Jarmon, an old-growth champion introduced in earlier chapters, presents this possibility when critiquing scientific orientations towards the natural world. In search of an alternative vision, Jarmon surmises that "Native Americans lived in harmony with the Earth because they did have an emotional-spiritual connection with it." They could, she adds, "sense when to do something or not do something rather than try to figure it out mathematically." Such claims, however stereotypic, are not the exclusive domain of environmental activists; indeed, they are widely held. Headland (1997, 605) offers a good example of this when quoting a 1992 *Time* magazine article that depicts a place that "time forgot" and a people (a Pygmy population in the northern Congo) "almost supernatural in their abilities – reading the faintest imprint at a glance ... and demonstrating a photographic memory for terrain."

Claims of ecological nobility tend to rest upon three assumptions, which, in turn, have profound social and political implications for those so stereotyped (Buege 1996).

- As per the dictates of stereotyping, ecological nobility is generally attributed to groups, not individuals. Claims assigned to all members of a group are essentialist in that they assume that all members of the group behave in accordance with that essence or feature. This sets up an impossible standard whereby anyone who does not then live up to that essence (e.g., being ecologically noble) falls from grace, or, in Buege's terms, "forfeits their ... nobility" (77). Moreover, when the romanticized "other" "fails to meet our expectations, we become judgmental of them" (83). If a Coast Salish person fishes with a powerboat rather than a spear or weir, he or she is dismissed as a "false or inauthentic Indian" who has failed to abide by the cultural criteria imposed by non-Aboriginal projections of ecological nobility.
- It is generally assumed that Aboriginal peoples maintain "a direct 'connection' to a particular environment" (74) and that to move is to lose that connection. This suggests that all contemporary and archaic Aboriginal peoples should have stayed in one place. Yet much in the modern and historical record confirms a propensity for peoples across time and place to travel, intermingle, and respond creatively and practically to encounters with "non-local" peoples (Clifford 1997; Marcus 1995; Wolf 1982).[9] Moreover, if we begin, stereotypically, with the assumption that the Aboriginal person ought not to move, and that, if she does, then she is somehow less Aboriginal or authentic, then two things become possible. First, those who do leave due to economic or political opportunity, marriage, or simple curiosity may come to be defined by non-Aboriginals as insufficiently Aboriginal and, thus, be threatened with the loss of treaty status (Buege 1996, 83). This is precisely what happened to the Mashpee people of Cape Cod's "Indian Town" (Clifford 1988), where the Mashpee had to prove continuous residence (a misrepresentation of Mashpee life) in order to prove they were truly Mashpee. At stake was the return of land previously lost in the courts. In other contexts, the problem of residence versus non-residence, or even traditional land use versus non-traditional land use, sets up an impossible quandary: "If [Aboriginal people] fail to exploit their land and its resources, they face threats that Euro-Americans will follow traditional Euro-American patterns and find ways to exploit these resources themselves. If Native people do exploit their lands, they risk losing their authenticity in the eyes of Euro-Americans" (Buege 1996, 86).
- Finally, the stereotype assumes that the connection to land "inherent" in all Aboriginal peoples has to do with inviolable respect. The implied ethical

order places environmental concerns and protection above all other human and social needs. Such assumptions neglect the very real needs of many contemporary Aboriginal people; it denies recognition of the fact that many cannot afford to pursue such idealized goals because their basic need for employment income, food, healthcare, and/or adequate shelter remains unmet.

Of course, asserting that ecological nobility claims *should* be challenged does not make it so in civic life. The picture is considerably more disordered. Many ancient forest activists do succumb to stereotypic references to Aboriginal peoples and past epochs generally. But many also conscientiously check their behaviour and are thereafter quick to qualify their claims. In one breath, someone might say or wonder whether the past was glorious and invoke the possibility of social change ordered on principles derived from small-scale societies. In another breath, he or she might counter these assumptions and convey considerable discomfort with discussions of Aboriginal-inspired "ideal" cultural arrangements. A similar confusion is evident in loggers' references to past and Aboriginal societies. A closer look at this disordered pattern reveals that activists on both sides are working to reconcile widespread nostalgia for the past, and the respect for Aboriginal land-use traditions that such nostalgia implies, with the possibility that loggers' claim to cultural recognition of their land-based communities carries with it a similar respect and thus the right to decide upon future land-use practices.[10]

Nostalgia Tensions

The nostalgia of some environmentalists, and its accompanying glorification of past epochs, is rooted in a profound sense of loss and regret for a physical world they imagine but can no longer experience. This is most commonly expressed through a wish to see or visit a preindustrial period with which Native Americans are closely associated. That envisioned preindustrial world also assumes a quietude; preindustry suggests a land where people are neither working nor living actively but, instead, are moving about quietly and inconspicuously (White 1996). Even a pragmatic, long-term, and mildly cynical activist like the past director of Ancient Forest Grassroots Alliance (AFGA), Paul Wilson, an informant who rarely spoke of anything but forest policy and forest science, indulged these imaginings. One weekday afternoon, as Wilson and I sat over coffee in a largely vacant Eugene café, he suspended his rapid-fire monologue on Forest Service corruption and resultant overcutting and said, in a hushed voice: "I mean to walk across North America in its natural state [pause], that would be incredible ... There's a lot of places I'd like to be but North America ... when North America was intact. I couldn't resist that."[11]

Forest ecologists are quick to point out that a baseline pristine past is the product of our imaginations, that change in biotic systems is constant and that no forest is solely natural. There is no point in time wherein one might find an unaltered past land against which all subsequent landscapes are judged.[12] But the point for most ancient forest activists and conservation biologists is large-scale anthropocentrically driven change and the inexorable link between a pristine past and a pristine people. This feeds a nagging desire to know and to pursue the question of how far back (and to what social circumstances) one must reach in order to capture a pristine ideal that might serve as a template for new ecologically sustainable possibilities. For David Burns, the appropriate metaphor for this persistent question is a time machine. "I have this time machine fantasy where I'd like to stand on this mountain and go [snaps his fingers] a hundred years ago, [snaps his fingers again] five hundred years ago, ten thousand years ago. You know, I just want to see how it works ... I want to see how it all works together and what we've done to it. I'd like to see a pristine world and that would be, you know, a pretechnology, preindustrial society." But, as noted above, not all references to the past are as unabashedly romantic or wistful. Julie Dawson finds herself admitting that she has a "warped view" of the past, whereas Greg Norman admits that "most of the time ... I'm really conscious that I've romanticized [the past]."

These self-criticisms are extended equally to their activist colleagues, particularly as they concern behaviour seen as disrespectful of Native Americans. The "use" of Aboriginal activists at conferences is often taken as a case in point. Early one spring, I sat in a different café in a different Oregon city, talking with peace and environmental activist Monica Ladner about the motivations that propel forest activism. Monica observed that the most enduring and effective activists in her group are generally committed to the physical areas they work to defend, but she described herself as resenting activists who "tack onto ... Native American spirituality." "I have some concern," she says, "about [environmentalists] taking on this pseudo-Native American covering." Particularly disturbing to her were her male cohorts' unreflexive appropriations of Aboriginal spokespersons at conferences. She recognized the symbolic capital inherent in the contiguous presence on the conference stage of environmentalists and Aboriginal activists, but regarded such collaborations as largely shallow.[13]

What I see is that it's White men trying to tap into something here without really delving into what it means, or why it is the way it is. What I've found and this irritates me no end, is they'll invite a Native American person to d an opening ceremony, and then be standing behind the scenes looking their watch about how long it takes. Or scheduling the Native Ameri speaker into a panel at the end so people can leave if they didn't want to hear everything that this person had to say.

There was also a concerted effort on the part of some activists to specify and contextualize their references to Aboriginal people, to reference "an" Aboriginal person and his or her nation rather than "all" Aboriginal people, and to do so knowingly. When I asked Andrew Simon about the tendency among some activists to attribute an innate ecological nobility to all Aboriginal people, he stated that he "want[ed] to handle that [subject] carefully because he want[ed] neither to lead the unknowing down the wrong trail nor to offend tribal people."

On the Problem of Cultural Legitimacy

In the end, evidence for stereotypic and romantic references to past peoples is available, as is evidence for some self- and inner-group monitoring among environmentalists regarding their relationship to such practices. This variable pattern is further complicated by the tension between the thirst of some environmentalists for a utopic future rooted in a simpler time and their tendency to dismiss rural logging communities as dystopic, as having nothing to contribute to that future. This utopic-dystopic tension is an outgrowth of the discomfort of some environmentalists with the possibility that their rural, timber-activist opponents possess their own symbolic capital – a capital sustained by lengthy association with the rural west and the idea that rural living is more emblematic of an ecologically benign lifestyle than is urban living. Consequently, environmentalists' glorifying discourse about past and other cultures (usually Aboriginal ones) is plagued by (1) the ambivalence many green activists feel towards the logging towns that populate rural Oregon and (2) the possibility that, by dint of history, loggers have a greater claim to place than do urban environmentalists.

The problem for environmentalists is: Under what temporal, social, economic, and even spiritual arrangements do claims of place attachment become legitimate? Or, under what arrangements do people become not simply temporary residents of a place but culturally recognized inhabitants of a place? We know from Chapter 4 that environmentalists can and do speak disparagingly about "diseased" and "dysfunctional" rural communities. And yet some also express, in the words of southern Oregon activist Nathan Carver, a longing for rural-like qualities in their own lives: "I long for the experience of living in a community and in a community that has a distinct sense of place, a community which includes a whole range of ages and that somehow has learned to honour and take care of all components of itself." Consciously and unconsciously, activists are sorting out what does and should constitute a healthy community or culture.

A good example of environmentalists' longing for community and their concomitant antipathy towards logging communities is offered by Julie Dawson, the mother, editor, and environmental activist whose wariness about "having science on [their] side" was considered in the previous chapter. During one of our meetings, Dawson recalled a vacation foray to a rural Oregon town of fewer than 1,000 people. She was struck by her own anomalous observations of rural Oregon life while sitting in the town's bar.

> When I go someplace in [rural] Oregon, as I did a couple of weeks ago, and just hung out in a bar with a friend and saw the culture that really exists there, I can have some sympathy for losing that. Because there really is something important about rural culture that we don't have. And I can see if the environmentalists just totally win this whole controversy and we move away from rural economies here and start getting into tourism like California and Colorado – I mean, we're just going to really change our whole state, basically. And I think that'll really be a big loss.

Yet shortly thereafter I asked Dawson if her appreciation of the quality of life in rural Oregon means that she agrees with loggers' claim that their struggle is for cultural survival and that, as the slogan goes, "people count, too." Dawson caught herself, indicating a clear discomfort with any conceding of ground on the implicit debate about granting loggers cultural legitimacy on the basis of multigenerational, rural, and land-based living. She was careful to refer to loggers as upholding a "lifestyle," an ostensibly lesser entity than a rural culture per se. She also delved into an impatient criticism of those who defend the cultural and individual standing of loggers.

> I just get really sick of hearing that. I've heard that so many times: "People count too. People count too." I mean I guess I can see where it's coming from but it's just nauseating to me. We've destroyed 95 percent of our native ecosystems. People are encroaching everywhere, in every other animal's habitat, and driving more and more species every day to the brink of extinction. "People count too" is just ridiculous.
>
> So when I hear the political whining and stuff about attachment to lifestyle and whining about cultural [loss], I think it's just pathetic. They have no argument. It's only the last resort of people who've become irrelevant, economically. You know, they have no argument. I guess that argument does not go really far with me.

For the most part, Dawson is frustrated by the emphasis on human priorities at the expense of the non-human world. She also asserts that loggers' claim to place is nothing but a "pathetic attachment to an economically irrelevant lifestyle." This helps her to resist loggers' efforts to assert their cultural status. Their claim to place is converted to base self-interest, to an economic plea masquerading as cultural survival.

Cultural Inauthenticity and the Unrootedness of Loggers

Given the importance of authentic attachment to place to the debate between loggers and environmentalists,[14] we can expect others to follow Dawson's lead by contesting loggers' status as physically and, thus, culturally attached to rural Oregon. Such examples are readily available. Among the most interesting is the effort to assert loggers' lack of cultural authenticity by portraying logging communities as unsympathetic and diasporic,[15] as rootless (i.e., place-unattached) and/or mobile communities of people who are removed or willingly distant from their geographic homeland. Initially, the term "diaspora" referred to the dispersion of Jewish people into Gentile nations, but it has since been extended to refer to refugee populations, immigrants, expatriates, and others (Clifford 1997; Tölölian 1991). Movement, or, borrowing Clifford's (1997) term, "unrootedness," can be the consequence of war, natural disaster, or the yearning for better opportunities. But because these diasporic experiences are generally regarded with compassion, and because such compassion would undermine environmentalists' attempts to disallow loggers' claim to forests, it is strategically necessary for environmentalists to assert that unrootedness in an unsympathetic fashion.

This is deftly accomplished by Matthew Waite, the high-profile activist introduced in earlier chapters. Waite had been noting how loggers misread the motives of environmentalists. He was telling me that loggers resist or ignore as incomprehensible the premise that environmentalists want to save old growth for the wildlife that inhabit it. He defined his opponents as unwilling to accept at face value the environmentalist justification for old growth: that old-growth forests are simply necessary habitat for the spotted owl. Instead, noted Waite, environmentalists are regarded as urbanites in search of rural playgrounds for their own gratification and for being "enviromeddlers" bent on "disrupting their [loggers'] lifestyle."[16] Waite followed this with a theory of why this anger about "lifestyle disruption" might occur and just what kind of lifestyle has indeed been disrupted. His explanation is gleaned from Social Darwinian theories of cultural evolution and contains a twofold assumption.

First, Waite refers to present-day loggers as unfit remnants, or "incorrigible cases," left over from a broader, more vital population. Loggers, for Waite, are stubborn contrarians who have resisted larger demographic patterns such as the post-Second World War move towards urbias and suburbias. These "people," notes Waite, "[are the] brothers and sisters [of those who] heeded the pull of what's going on in this country – from the rural to the urban. Their aunts and uncles went off to the city, their great aunts and uncles went off. But this is [a] generation ... of the *self-selected*, people who are *terrorized* by the thought of the city." Waite's comments then turn quickly to a discussion of his own family; he is in search of an example of a more appropriate evolution, a "progression" from the labour of logging to a more socially mature, upwardly mobile, and inherently class-enhancing lifestyle. Waite recalls that his grandfather went from "hauling logs with a team of mules" (which is, notably, a benign or ecologically low-impact means for extracting felled timber) to being a "custodial worker at the local high school." His mother graduated from college and, one is left to assume, the family's fate improved from there.

Second, Waite regards loggers as a class of people who have no historic claim to the land (in this case forestland). Rather than representing a culture or way of life that is land-based, loggers, for Waite, are rootless pools of labourers who, by quirk of circumstance (the circumstance being the sheer abundance of forest in the Pacific Northwest), were able to settle for more than a single generation. "The thing about logging," asserts Waite, "is that only in this forest, because [it was] so large and so vast, did loggers set roots. They were always nomads before. They would cut for a generation and if [their] kids went into the business they went somewhere else. Their roots were not established." Waite compares loggers to miners who, he says, "don't live there [at the mine site] forever."

The characterization of loggers as a transient labour pool and the concurrent undermining of their claim to a place-rooted lifestyle is further buttressed by Waite's ability to recast questions about culture into questions about the availability of employment. Thus even when I ask him, point blank, whether or not the forest dispute is indeed a cultural war, he says only that loggers "feel" it to be a cultural war. Thereafter he defines activist loggers as belonging to an opportunistic movement that attempts to convert an ordinary economic downturn into an act of cultural obliteration. Waite achieves this discursively by limiting the causal explanation for reduced logging opportunities to that of market forces. Despite the glaring presence of the federal court's ban on logging (as covered in Chapter 2), Waite offers as an analogous case only those instances in which job losses were driven by changes in consumer preferences and availability of alternative commodities.

> They are feeling it to be a cultural war ... and I've been accused of practising cultural genocide. But again, they're incorrectly analyzing environmentalists. Environmentalists don't want to destroy their lifestyle, but they don't put an extremely high value on it either because these guys say, "Well, I don't want to move" [mocking voice]. Well who here hasn't had to move for their jobs – 98 percent. And so it just kind of rings hollow to say, "I don't wanna." I mean it works well here, it sells well among themselves, they

believe their own stuff ... [but] what's going on is there are macroeconomic factors ... I mean, when the mining petered out [or] the market changed on beaver hats, was that destroying a lifestyle? Did the trappers say, "Oh god, you know, this is destroying our culture!"

Waite is right to suggest that we tend to treat job losses stemming from depressed market conditions or changing patterns of consumption as acts of god whereas policy-driven or advocate-driven changes are often perceived by the affected group as deliberate attacks. But this does not preclude the fact that Waite is casting loggers as atavistic (and thus unsympathetic) curiosities and as a rootless migrating labour force in search of a market. Loggers' rootlessness is established by Waite's suggestion that the Oregon context is an anomaly, to wit, a territory of dense conifer forests that has provided loggers with continuity of employment, if not "culture," across multiple generations. Their moral atavism, meanwhile, is suggested by Waite's assessment that loggers have failed to "progress" towards urbanism and middle-class professionalism.

Becoming "Indian"/Building Authenticity

The centrality of cultural authenticity to environmentalist-logger debates is also manifest in the attempts of environmentalists to acquire their own place-centric legitimacy, including novel efforts to "establish" Aboriginal-like sustained commitments to physical locations.

Consider Andrew Simon's creative attempt to establish viable place traditions by writing his own history onto the land and forests nearest his home. In earlier chapters we learned that Simon came to forest activism through a circuitous route – from urban-intellectual to craftsperson-carpenter – only to find himself living in southern Oregon in the 1980s as part of an increasingly settled community of back-to-the-land (and, by definition, rural) aspirants. In the 1980s he became involved in a highly publicized battle over Forest Service efforts to build roads in a biologically important region of rare species of cedar and fir. One segment of the region, known as Sumner Mountain, was the subject of bitter confrontations between Forest Service personnel and environmentalists. The protests garnered public attention and greatly increased the profile of direct-action environmentalists in Oregon. Simon offered a sarcastic characterization of the period: the one in which the Forest Service sought to build roads "to prevent the disease of wilderness from spreading out any further into the forest." Simon's early contributions to the confrontation included acts of civil disobedience and protest, but he was quickly dissatisfied with his role as a "weekend warrior." Older by one or two decades than most of the protesters, Simon was "unmoved by the heroics of going to jail" and, thus, began searching for acts of disobedience that would offer something more than transient satisfactions and cyclic periods of political elation and exhaustion.

> All my life up until that point, my frustration was a feeling of being split, it was like I couldn't get my political values, my spiritual urges, and my survival needs to work in such a way that they supported each other. I sort of would get off one limb or the other, discover that I was then not taking care of some other vital part of my life, and scamper[ ] back and be a yo-yo. You know, intense political burnout involvement, burn out, drop out ... That first summer I realized that, "Gosh, here I am just sitting on this mountain and taking care of it, and it's a political statement." [It was] the first time in my life that I wasn't feeling those splits, those contradictions.

During this period, Simon noted that Aboriginal activist, author, and orator Warren Lyons had convinced him that, metaphorically, the death of the Earth, or biosphere, would be coterminous with the final and complete loss of Aboriginal cultures. The message received by Simon, as a member of Lyons's audience, was "that somehow ... when the last Indian was gone everything was gonna go right down the tubes ... [because] no one was speaking for the wild."

Without sounding facile, Simon reflected that, despite his activist efforts to date, he had "never even seen Sumner Mountain" and that somehow he must resolutely contrive to "establish a relationship with this place." What followed was a journey designed to address Simon's desire to conduct meaningful acts of social change while simultaneously addressing the problem posed by Lyons's speech: the need to repopulate the Earth with Aboriginal values. Simon set out to achieve this by embarking on a multi-year effort to develop a relationship with Sumner Mountain. In so doing, he metaphorically becomes an "Indian" in that he becomes someone who acquires authenticity (defined in this context as sustained dwelling in place) by virtue of his long-term commitment to Sumner Mountain. I do not mean to suggest that Simon grants himself tribal prestige or title: he would be the first to resist such claims. What is clear is that he is one of many activists trying to embody practices that refer to and directly reflect White understandings of Aboriginal cultural arrangements.

For Simon, enacting new possibilities for the human use of and posture towards Sumner Mountain involved making the transition from visitor to resident. He set out with the intention of staying for five days only to have that effort evolve into an annual, sustained practice.

> I went in with five days' food and I ended up spending fifty-six days because, as various people came along and I told the story [of his Lyons-inspired route to Sumner Mountain], they would turn around and bring more food

in for me and so forth. In the process I sort of got married to a place, you know, got backed into a marriage with a mountain. I had no idea how to deal with this, and I simply said, "Well, I've got to empower this feeling somehow." And I thought I'd create a little sanctuary on top of the mountain. I had cleaned it up; it had been an old fire station lookout. Big garbage dump up there. And again, remembering a book that I had read that described the medicine wheel, and I didn't consider myself on a Native American path or anything, but the idea of talking to the directions and the four winds and so forth, was just perfect language for a mountain top, and I said, "Okay, I can make a circle," you know, so forth and so on. When I'd completed it, I really liked it, and I simply made what ended up to be sort of a marriage vow. I said, "Well, what I'll do is I'll come back every summer and just take care of it, keep cleaning the place up, be of service if I can and make walking sticks." I gave them away to people.

Simon is fully cognizant of the derision that his creative acts of resistance inspire in those he knows, including his more cynical, "rational," and/or urban-centric colleagues. He recounted their reactions by way of a mock dialogue between himself and his skeptical acquaintances.

People say to me:
"Andrew, come on, you have a relationship with trees?"
And I say, "Well, yeah."
"Oh come on, how is that possible? You mean like talking with trees and stuff?"
And I say, "Yeah."
"How is that possible?"
And I say, "Well, you know, you're not thinking about it right. If you were thinking of another human being, if you want to have a relationship with that person, what would you do?"
And they would say, "Well, I would hang out with that person."
And I say, "Well, ultimately could you explain your relationship in terms of what was literally said?"
And they say, "No, it's about something more, it's about hanging out together for a long time."
I say, "Okay. That's what I do with a tree or with a place. I hang out with it." My notion is this: There's a difference between comprehension, which is a mental thing ... and the word called understanding, [which] literally means to stand under. In other words, to simply place yourself at the foundation of.

Simon's annual dwelling at Sumner Mountain and his persistent efforts to render sensible the development of relational interactions with the mountain provided him with a modicum of local notoriety. His actions can also be seen as a manifestation of environmentalists' larger effort to establish claims to land that regard the physical world as something more than a commodity to be extracted for trade or profit. For better or (from the loggers' perspective) for worse, Simon has become a symbol of the effort to undo the stereotype of environmentalists as urban intellectuals devoid of experience within, or claim to, the physical world they seek to defend. Through residence on the mountain, he has sought to "earn" for himself – and, by extension, other members in the local environmental movement – something akin to "place authenticity," which, in turn, confers the right to speak for and defend the disputed physical territory. Moreover, the use of Sumner Mountain as a refuge (as opposed to a land base for the provision of food) and Simon's attempt (albeit limited) to mark his place by appropriating the "medicine wheel" practices of "traditional" peoples, implicitly assigns the mountain status as sacred terrain.

Imagined Futures

Romantic essentialism refers to the practice of valorizing as noble those cultural communities that White activists define as authentic. Strategic essentialism occurs when Aboriginal or other subjugated peoples amplify, in their dress or discourse, aspects of their culture that conform to White notions of nobility; they recognize the political potency of such behaviour and the benefits that can accrue from it (Brosius 1999a). A ready example can be found in Conklin's (1997) work on Kayapo use of traditional dress at rain forest conservation conferences attended by White European and American activists (whose support they pursue) (Conklin 1997). The Oregon context further complicates this distinction. Most ancient-forest activists are not themselves subjugated people, but they often defend those who are.[17] They practise romantic forms of essentialism (described above), and they also use their alliances with, and emulations of, Aboriginal people to their own strategic ends.

Yet dismissing environmentalists as solely romantic or strategic is insufficient to the extent that it silences their endeavours to think deeply about Aboriginal practices and to reflect upon them when attempting to imagine an ecologically viable future. A better approach is to examine these imaginative flights of fancy so as to understand how their speculations both intersect with, and are constrained by, references to "natural" land-based communities. Among the many problems environmentalists attempt to address, four are paramount: (1) problems of scale, (2) concern about nature-dominating systems of thought within modernist intellectual traditions, (3) the problem of containing human misbehaviour towards the environment, and (4) the question of what knowledge base one should refer to when deciding how to act or how to become a "good" land manager.

The question of scale emerges when activists grapple with the concept of ecological sustainability and/or the Earth's carrying capacity. Points about scale also emerge because it is difficult for any one person to imagine a reorganization of human and natural systems on a global level. Popular though globalizing "gaia-like" discourses are, activists generally relied upon examples of small-scale agricultural and/or hunting-gathering economies in their efforts to articulate new practices. Activist Jim Border, then volunteer member and president-elect of AFGA, speculated briefly about ideal human-nature relationships.

> I think that we have to live with it [nature] instead of fighting it all the time. Poisoning it, killing it. Our survival depends on it. The human race cannot go around killing everything. Take the Native Americans, [they] had a great environmental ethic. When they got cooped up in one place, they moved. And there were few enough people that it [nature] recovered. We didn't have the capabilities of doing the massive destruction we do today, nor was there the amount of people. And they were being careful with the destruction that they do.
>
> They killed animals, thanked [them] for [their] contribution to their well-being, but they didn't hesitate to kill [them] or anything. But they weren't in the numbers to cause extinction either. They, as people, felt very close to nature. They didn't have the manifest destiny and all these bullshit things that the White people dreamed up.

The problem of scale asserts itself in this passage as Border surmises that past environments were healthier due to lesser population pressures. But his speculations are also confusing; he oscillates between assumptions about ethics driving behaviour and the possibility that the comparatively sparse (and, one assumes, technologically minimalist) populations of the precolonial era were incapable of inflicting destruction on a vast landscape. Such ambivalent assertions, along with tangled webs of assertions and counter-assertions, are a by-product of the unsupportable standards for behaviour incurred by romantic essentialism. Speech and thought are trapped, à la Buege (1996), by reductionist and imaginatively limiting claims.

The inconsistency of thought reflected by this passage appears less odd, however, if one focuses not on the speaker's indecision about determinist (ideational versus material) models of behaviour but, rather, on the messy sorting out of the future and past that engages many activists. Central to this sorting is the question of how nature might become the dominant organizing principal of society and what that would mean for human behaviour. Frequently this leads to speculation about human existence under "naturalistic" conditions, which are defined as conditions that situate humans as a species among others within a "functioning" ecosystem. To clarify this, one might borrow science writer David Quammen's (1998) "weed" metaphor. Some environmentalists imagine humans as a weed species that has invasively colonized the Earth, whereas the point about "new" cultural forms is: What does it mean for humans to be an indigenous rather than an endogenous (weed) species?

A good example of this speculation and the influence exercised by fears about overpopulation alongside assumptions about ecological nobility is found in activists' adherence to ecological-niche models of human culture. Michael Costas mused about this when recounting a recent trip to the Sinai. He first emphasized the comparatively recent expansion of the Sinai-Bedouin population, which he attributed to the introduction of modern medical care. For brevity's sake, I paraphrase Costas, preserving most of his original wording:

> Here was a culture based on grazing and the search for water, but also renowned for a poetic version of Arabic widely regarded as an art form. Historically the population stayed in balance because infant mortality was so high [and] the climate so harsh [that] the population stayed down. But once we introduced medical care, the population started expanding for the first time and now the ecosystem's completely breaking down. They're trying to resettle them on the coast, putting them in houses, and trying to introduce population control.

But Michael's own eco-noble, niche-centric preoccupation with how to limit population leads only to privileged musings about our "unnatural" paranoia about death.

> I think that a lot of times we shouldn't tinker ... we've got this ethic that death is so bad, not recognizing that this whole thing [niche] works on life and death and death is not so bad, that it's a natural part of life and it's okay. Everything's gonna die. It's okay, guys. And even though we've got this whole revulsion to it, I think we need to really seriously examine where that comes from. The emotional parts. My personal stuff, the idea of my kids dying scares the hell out of me. What a horrible, horrible thing.

The position that excessive human mortality is beneficial to ecological health has attracted considerable critical attention. I find, as do many others, the very idea of imposing high mortality rates on disadvantaged populations – under the guise of averting ecological crisis – a form of violent colonization of the first order. Though to Michael's credit, he attempts to envision the implications of this model for his own life, and appropriately imagines its consequences as "horrible."

The nagging question remains: What do these inhumane musings about alternative cultural forms offer environmentalists and why do they persist? One reason is that, ultimately, they contribute to the larger activist effort to play with and think through wild or outrageous scenarios of change in the search for workable ones. Human life is prescribed by existing cultural forms – in Michael's case, by dominant discourses and assumptions about noble Aboriginal practices as the antithesis of an overly populous, overly consumptive human presence on the Earth. Michael is working, consciously or not, with that form or popular assumption in order to cultivate change, using it as the baseline from which innovation begins.

One can see, in the discourse of activists, an emerging spectrum of possibilities, each premised on the same dominant "ecologically noble" narrative with which Michael works but increasingly devoid of the inhumane and naive qualities that trap one within an impossible loop – torn between the consequence of overpopulation and its genocidal opposite. They embrace Michael's speculations about changing cultural forms, but, rather than thinking about how to "return" to "living in harmony with nature," they question how society might cultivate practices that (1) mimic natural processes and (2) restrict ecologically deleterious behaviour.

Jeff Milton, a former Forest Service employee and AFGA activist, offered an example of mimicking "natural processes" when promoting the use of fire in land management. Fire is associated with both Aboriginal practice and natural process and has recently been elevated to an ideal practice (Pyne 1997). As a powerful and often uncontrollable force, fire has achieved status (conceptually at least) as a form of land management that counters modernist hubris about human ability to control nature. In other words, fire is a perfect metaphor for nature "returning" to its status as the dominant force in culture and for management efforts that embellish rather than suppress naturally occurring disturbance regimes. Milton becomes extremely animated when suggesting its possibilities:

> I wish we knew what those folks that came before us, how they used fire and what they did ... I really think that the way we're going to finally learn to manage the forest is to admit that we don't know what we're doing, and to admit that the way it's happened naturally for the last at least 9,000 years, since the last ice age, is probably the best way to manage. Once we decide that natural disturbance is the template that we should use, then I think we'll be okay.

Andrew Simon concocted similar innovations based loosely upon Aboriginal practices but focused upon the need to restrict anthropocentric degradation. Note, however, that he does not expect his ideal human community to "live up to," or embody, nobility. "I'm more sanguine," says Simon, "about changing human culture than changing human behaviour. If I thought that we had to change human nature, I think I might get really despairing." His speculations about future models for culture focus on dismissing "genocidal" imaginings, all the while accommodating the exigencies of human behaviour.

> Unless we're going to have a public policy of AIDS or something like that, which may end up [happening], but it ain't gonna be public policy, the question becomes ... how to describe this?
>
> Let's go to hunter/gatherer[s] for a minute. If you were a young, type A, achievement-oriented, aggressive, lustful young male ... you would get your status from being a good hunter. If you wanted to be a good hunter the last thing that you would want is a lot of possessions to carry around. Therefore you compete with your adversary in terms of how much you can give away. And you get the potlatch out of that. So suddenly, something called greed, lust – very human traits, right – become channeled into collective benefit and so forth, right? That's an economy. Alright? Now another one adds up with four cars in the garage, and all kinds of electronic toys in a multi-room house, you know, and closets full of stuff, and so forth and so on, which is another way to impress the girls, but results in a different outcome.

Simon's hunting-gathering example is not meant to be a precise and theoretically defensible portrait of past peoples; rather, it is an instrument that he uses to create an "as-if" social world, a world he hopes might eventually come to pass.[18] The Aboriginal-inspired illustration allows him to address the question that most preoccupies his musings: How is it that we can describe and think about new possibilities; how can the future be rendered imaginable and thus possible? "I'm not trying," Simon offered, "to romanticize, in any sense, the hunter-gatherer life, because it had the full [gamut of human problems]." But he does admit to "reaching for a storyline" that addresses "how we can be here for a long time."

In so doing, Simon appears to draw upon philosopher-forester Aldo Leopold's conservationist motto: "The first rule of intelligent thinking is to keep all the parts." He wants to "keep and use" the pieces of a healthy natural system, as well as those pieces of healthy cultural systems that will permit (or form the basis for imagining) what environmentalists typically refer to as survival.

> The central question for environmentalists is figuring out how civilized people, without having catastrophe throw them back into what you would do as a hunter-gatherer, [is to] figure out whether or not civilized people can write a new story for themselves, and enact it in such a way that they do in fact leave a future ... [or] evolve into new ways of being, which will be a slow process ... In this period, two arcs need to be preserved. One arc is any wild places [where] you have the gene pool. The other arc [is] the stories and traditions of Indigenous people, because in industrial culture we don't have the storyline of how to live in a place long and well.

The final point in Andrew Simon's speculations picks up a thread that was first introduced in Chapter 5; that is, the question of choosing a knowledge base worthy of informing forest management. The "absence of a storyline" suggests to Simon that the alternative to the spurious and ill-informed science of ecosystem management eschewed by environmentalists is something he might call deep tradition. In particular, he is interested in traditions that respect the knowledge born of living in a place that mimics natural cycles of decay and recovery, and that offer systems of thought that normalize the idea that humans are not conquerors of natural systems but, rather, are vulnerable to them. To illustrate this possibility, Simon retells a story provided to him by a college student who came to visit him during one of his above-mentioned sojourns on Sumner Mountain. The story, said Simon, helped him to reconcile himself to the destructive might of the forest fires he had witnessed the previous summer.

> A college student came up to visit, he was in Portland. The previous year he had done fieldwork among the Masai. Due to a happy set of circumstances that were actually quite unusual, he got really tight with one Masai warrior and was invited to participate in some events that no one else was invited to. Early one morning, well before dawn, this Masai warrior showed up at his camp and says, "Come on." And he went on a walk with four others for a long time. Just about dawn they came into a deserted village, just huts, no sign of life or anything. They sat down in the centre of the village, and they started to drum. Slowly. What they were doing is they were marking the end of a one-year mourning period that followed the seven-year drought. In Masai culture, one accommodates to nature's cycles and expects that once every seven years there will be a loss of a significant number of your cattle and several of your loved ones. In the aftermath of this death experience, they go into a period of mourning in which there is literally no village social life for one year. People stay in their huts, they come out only for survival reasons and body functions and so forth. They sat down, they began drumming, and they were drumming the village back to life. And across the next twenty-four hours, this built into a total orgy of new life. And this is repeated once every seven years. Now the modern civilized technologist would look at this and say, "That's stupid, the answer is irrigation and birth control." On the other hand, I'm not quite sure that those people haven't successfully figured out a way of living with their place, all right?

Nobility and Anti-Nobility Discourses among Loggers

The proposition that references to Aboriginal practices reflect the importance of cultural authenticity to activists' imaginings of future land-use practices can be clarified by a parallel examination of loggers' arguments. Like environmentalists, loggers are also in search of an authenticity-based "right" to determine Oregon's forest practices. Virtually all of their references to Aboriginal peoples involve efforts to either deconstruct the ecologically noble stereotype or to cast themselves as analogous to them and, thus, as deserving of a comparable level of respect. Upon perusing these findings, and juxtaposing them to what has already been said about environmentalist positions, it becomes patently clear that Aboriginal peoples are caught up in a pro/anti dialogue not of their own making – a dialogue between competing White activists (loggers and environmentalists) that, at times, is ugly, racist, and dehumanizing. It is a dialogue that reduces Aboriginal peoples to unidimensional strategic and counter-strategic implements caught in the crossfire of the struggle over Oregon's forests.

Denouncing Ecological Nobility

Consider first how loggers denounced the ecological nobility stereotype, as this took several forms. One was to accuse environmentalists of eco-misanthropy, of maintaining that humans matter less than do other species. Such claims were introduced as serious points of contention but also with the dry wit for which loggers are well known. Thus, one woman commented to me that she need not "see a psychiatrist because she cannot walk her dinosaur," implying that she need not secure her psychological well-being by caring for an extinct species. Accusations of eco-misanthropy are frequently presented in association with environmentalists' exaltation of imagined Aboriginal practices. A fitting example is provided by Bill Walton, who spoke at an annual gathering of Oregon's gyppo loggers. Walton is a motivational speaker seconded by the Wise Use movement to rally loggers with his hyperbolic presentational style and his story of metamorphosing from environmentalist to extreme skeptic. His words should be read with this in mind, though much of what he has to say is supported by loggers with whom I spoke.

A recluse himself, Walton once lived on a remote island in the Great Lakes region nearest the Minnesota border. His talk opens with the suggestion that he and his wife lived a singularly "primitive" lifestyle: "We had none of the modern amenities that you would recognize as being a part of civilization. We had no electricity except for generators and had to go eight miles by boat or snowmobile to the nearest payphone." During this period, Walton recounts that he became involved with a local environmental group only to become the focal point of criticism for not living up to its standards of a "truly primitive" lifestyle: "When they [referring to environmentalists] found out that we cut our own wood with a chain saw, they resented that. They wanted me to go out and chop wood to carry us through the winter with a buck-saw and an axe. They resented the fact that we had a motorboat to get the eight miles up this big enormous lake. They wanted us to travel by canoe." Eventually, reports Walton, he became the nemesis of the environmentalists when he resisted their effort to include his land in a proposed 14,000-acre (5,670-hectare) wilderness area. Walton concluded his story by accusing environmentalists of wanting to "bomb us back into the stone age where we might live guilt-free at last." He then proposed an inflammatory mock plan to "give them [environmentalists] a ticket to Borneo or the Philippine jungle where there are people that live in the stone age ... See how long it takes them to plead for a return trip."

Though likely an apocryphal story, Walton's central rhetorical ploy is to parody the environmentalists' fascination with the lifeways of technologically minimalist peoples living in sparsely populated regions and/or precontact epochs. His conversion-story logic depends upon him being forced to forego his own chosen ecologically sensitive lifestyle because it failed to pass a rigorous "authenticity" litmus test. This implies that environmentalists are misanthropists interested only in those social possibilities that reduce human presence to postapocalyptic hunter-gatherer economies.[19]

Undermining the Symbolic Capital of an "Aboriginal" Presence

As common as accusations of misanthropy and reversion to stoneagism are, so too are references, following Berkhofer (1978), to socially and ecologically barbaric Aboriginal practices. These references showcase loggers' impatience with conceptions of post-1800s Oregon as having somehow fallen from an Aboriginal golden age. The goal, for loggers, is to undermine the symbolic capital that environmentalists accrue through their noble-centric invocations of Aboriginal pasts. Invariably, they involve denouncing Aboriginal people for living an existence permeated by discord, disease, and any number of social ills. "The whole thing about pre-Columbus time, those people [referring to environmentalists], in their mind everything that's bad about the world is the White European influence on everything. Everything presettler was good and everything postsettler was bad. If you read accounts of what the Indians used to do to themselves, it's not pretty. Everything wasn't a bed of roses for those people. They were at war all the time with each other."

Logger Steve Fuller's parodying of a "good past" versus a "bad postcolonization present" is also endemic to Mark Evans's charge that environmentalists are ignorant of what he considers to be Aboriginal peoples' ecologically destructive practices. Unlike Jeff Milton's earlier enchantment with Aboriginal uses of fire, Evans suggests that Aboriginal populations used fire excessively and thus left early Willamette Valley settlers with a degraded landscape. He portrays environmentalists as imagining that White explorers Lewis and Clark found an untouched paradise inland from the mouth of the Columbia River.[20] "People have the notion that when Lewis and Clark came down the Columbia they found a thousand-year-old forest stretching from Northern California to Southern British Columbia." Evans, a student of history himself, is not simply upset because his opponents are historically naive, but also because their assumption about a paradisiacal precontact Pacific Northwest carries with it the supposition that those living at that time left little or no imprint on their physical surroundings.[21] In order to counteract this position, Evans asserts the following:

> The Indians, when they crossed the Bering Land bridge and started down [south], they started burning it [the forest]. There's a gal who wrote a paper about the Northwest Indians. Out of 800 words, they had thirty that denoted fire. It must have been really important to them you know. They burnt the Willamette Valley so it wouldn't grow timber any more. It's kind of like what the Indians did with the buffalo in the Plains days. They might have wanted fifty or sixty buffalo to get them through the winter, or 500 buffalo to get them through the winter, but they drove 5,000 over the cliff. These guys would wait until the conditions were right or wrong and these guys might want to burn 5,000 acres and instead they got 50,000.

Evans's point about fire calls into question environmentalists' enthusiasm for land-management policies based upon natural processes and (especially fire-based) disturbance regimes. To some extent his comments also parallel the historical record, at least insofar as Aboriginal use of fire to clear forested areas and to improve hunting and gathering opportunities is concerned. Burning encouraged the appearance of coveted species of berries and deer in cleared areas (Boag 1992; Cronon 1989; Pyne 1997; Robbins 1997).

Beyond this, there is little certainty among historians about whether Aboriginal populations introduced excessive and/or catastrophic burning. They agree only that "the form and magnitude of environmental modification [and] whether Indians lived in harmony with nature" is an important question (Deneven, quoted in Robbins 1997; see also Booth 1994). The possibility of extensive buffalo kills is also debated, as is the importance of Edenic narratives wherein an imagined pre-White wilderness serves to inculcate the need for mainstream Whites to rediscover the benefits of nature and wilderness per se (Slater 1996).

As if to mirror these intradisciplinary debates, Evans also expresses his own ambivalence about the imagined pasts of Aboriginal peoples. On the one hand, he actively discredits popular notions of Aboriginal peoples as incautious about the well-being of their animate and inanimate surroundings. On the other hand, he has an awe-like fascination with their extensive vocabulary for (and likely knowledge of) fire. This is best evidenced by his juxtaposition of paradoxical images: buffalo slaughter and excessive burning versus the [traditional] knowledge implied by many words for fire. Possibly, Evans is himself unsure about just how far to extend his implicit critique of the nobility thesis.

Nobility as Performance

For loggers, a third category of objections to the nobility thesis involves viewing Aboriginal-environmentalist alliances as inauthentic and inaccurate performances. Stacey and Cheryl Mott offer a case in point.

When we met at their home after dinner one evening, Stacey was still shaken by the recent loss of his night shift job at one of the town's three mills. Cheryl continued to work as a secretary for the same mill. Both were active with OFCC. Preceding the exchange that follows, Cheryl, Stacey, and I had been discussing their belief that only contemporaneous rural and local people have a legitimate claim to the region's old-growth forests. Within this conversational context, talk turned to recent protests at Pine Hills, in a nearby national forest, where protesters had gathered to stop an old-growth timber sale. Unusually for Oregon, Aboriginal activists had joined the protests because the sale trespassed on Chinook burial grounds. Stacey, however, used the event as an opportunity to articulate both his antipathy towards out-of-state (e.g., Washington, DC) people staking claim to Oregon's forests and towards an alleged collusion between "the media" (heretofore dismissed by Stacey as unabashedly "preservationist") and environmentalists.

> And I'll tell you another thing too, speaking of being out-of-state. These Indians, you know, that have been saying, "Oh, the trees, we worship the trees, it's part of our religion." Well some of them are from, you know, are from Arizona. And some of them are from the Dakotas, aren't they? We were talking about that at our last [OFCC] meeting. Ian knows, he researched several of them and told me the percentages of who they were and where they were from. Some of them were Sioux, I recall, Sioux from the Midwest. So there were multiple tribes, but people look at TV and they see someone that looks like an Indian, and they say, "Oh, surely they must be from here."

Stacey's mocking tone speaks to the tenacity and desperate single-mindedness with which some timber-community activists assert that claim to place is validated solely on the basis of one's geographic (and comparatively ahistorical) proximity to the territory in question. His repudiation of Aboriginal claims to Pine Hills denies, of course, the multicentury presence in the area of the Columbia River-based Chinook, their Calapooia neighbours, and the Coast Salish and Northwest Coast groups in the adjacent areas. It also denies the possibility of an Aboriginal "community" defined by something other than geographic affiliation or common dwelling. More important, for the purposes of this chapter, it is an anti-authenticity claim built upon White assumptions that first peoples may only act to protect territory when they can also "prove" a continuous presence in, or affiliation to, the place in question.

Much of Cheryl's distaste for non-local participants is assigned to those presumably non-Aboriginal persons (i.e., the "New Yorkers" in the passage below) who protest on behalf of "our" forests. Her forests are, legally speaking, national forests, but her conclusion is logical given the historical precedent set by the Forest Service, which, for decades, has contracted cutting rights and promised a "sustainable yield" to local people and industry. "We object to people who come from New York or even some of them from outside the United States, and protest in our forests if they're part of the national forests. But they're out of their territory. You know, I wouldn't go to New York and protest. I mean, I would feel stupid."

Loggers' critiques are also founded on the assumption that an authentic performance must entail the presence of a "traditional-looking" primitive person located in "pristine-looking" environments. An example of this is provided by Arlene Baker as we sit at her kitchen table with her husband and adult son. Father and son work for two different local mills; Arlene is employed as a clerk in a local hardware store. All three are active members of Caledon's group, Save Our Community. Referring to a recent television program advocating the protection of a tropical rain forest, Arlene offers a frame-by-frame analysis of the TV-mediated representation of the region's long-term Aboriginal inhabitants. With a conspiratorial nod, she tells me that her friend and activist colleague taught her to scrutinize the consistency of documentaries about environmental protection. Her critical retelling of one program included derogatory comments about Aboriginal persons as "nomads" dressed in "diapers." I do not edit Arlene's comments, for such oversanitizing acts conceal the caustic barbs that permeated the dispute. I do, however, assume that her vindictiveness is born, in part, of the fact that loggers bear the brunt of forestry policies not of their own making.[22]

Arlene begins her media analysis by noting: "First they're showing this man and his unique skills as a hunter ... these are nomads. They don't want to lose their rain forest and all the food in it. They're showing them with little darts that they blow and kill the birds with. And I thought, 'Wow, that's really neat.'" But later, as the program's content shifts to the ramifications of community encroachment introduced by resource extraction, Arlene points out that the rain forest setting has also been altered so as to pair the depletion of bird life with a depleted-habitat message: "All of a sudden, there were no birds. Where did they go? They were just totally gone."

Still more of Arlene's attention is captured by the inconsistent dress of a people portrayed as unaffected by the outside world. "What was really unique to me was they've never been around people. They've never let people in." Arlene is quick to represent herself as at first deceived but, upon recovering, of being invulnerable to representations of "authentic" people. "They were showing him a watch, [a man Arlene refers to as a chief] and he said, *Oh no, he didn't know what it was*." Critiquing his performance, she then adds that "he was dressed for the part." But his clothing, reports Arlene, directly contradicted the impression provided by the garments worn by those around him.

> He had [on], you know, I call it a diaper. No shoes. But the rest of the nomads had American clothing on. These are people who've never had any outside intervention. But there they were, you know, in American shorts and T-shirts and thongs. I said to my friend, "You have taught me that much." I picked up on it right away. I just couldn't believe it.

Arlene's selection of an Amazonian example to make her "deconstructive" point about the admixture of pro-Aboriginal and pro-environmental activist movements is fathomable because, following Slater (1996, 114), "few places conjure up similarly powerful images – great snakes, immense rivers, Indians with feather halos, and above them all a glittering canopy of green."

On the Implications of Nobility Discourses for Aboriginal Peoples

Loggers, for their part, must contend with the symbolic capital and political leverage environmentalists achieve through their invoking of Aboriginal alliances and practices. They do so by defensively critiquing portraits of ecological nobility, all the while falling into the incipiently racist pattern that Buege (1996) warns against. That is, when Aboriginal people are regarded not as a people like and unlike any other but, rather, as exemplars of an environmentally ideal cultural arrangement, then loggers (and others) tend to hold them to impossible "authenticity" standards and judge them harshly when they fail to meet them.[23] But Aboriginal/non-Aboriginal activist alliances have implications for both the Amazonian and Oregon contexts (Conklin 1997). Promoting, in the same political arena, Aboriginal and environmentalist causes invariably involves a reductionist portrait of the diversity of cultural and regional perspectives because the goal is to market "Indianness," broadly stated, and not diversity per se. The campaign's success depends upon the degree to which the projected images conform to narrow parameters prescribed by White definitions of authenticity; support is garnered precisely because mediated images play into Western notions of noble savagism. Semiotic devices such as visual exoticism (nudity, body paint, colourful ornamentation), as opposed to contemporary dress codes, are used to communicate the presence of an alien lifestyle, locale, or mindset to expectant Western audiences. Even though these acts can be entirely conscious and positively self-directed on the part of Aboriginal activists, they are paradoxical because Western notions of traditional culture, to paraphrase Conklin (1997, 729), insist that Aboriginal Amazonian activists embody authenticity (wear "costumes" long outdated, emphasize dormant customs, etc.); they are thus ironically forced to act inauthentically.

Within the context of the Pacific Northwest, the consequences extend further still. Unreasonable expectations of Aboriginal peoples set up by the ecologically noble savage stereotype intersect with an already pernicious competition over who can lay claim to land-use practices in the national forests. Environmentalists work hard to promote Aboriginal/non-Aboriginal alliances (in the case of the Pine Hills example), or the hint of such alliances (in the case of superficial Aboriginal participation at some conferences), because of their genuine interest in alternative cultural practices. But it is an alliance that also serves to counteract the insufficiently place-based claim of urban people to "rural" forests. By virtue of their association with Aboriginal activists (who, more than any single group, embody the idea of possessing a claim to non-urban territory) environmentalists shore up their limited claim to rural forestland and also creatively introduce their own Aboriginal-inspired experiments in order to further develop their claims to place. But the idealization of Aboriginal lifeways imposes expectations upon Aboriginal peoples concerning how they ought to use their own resources. Such has been the case with environmentalist criticisms of logging practices on reservation lands in the Pacific Northwest and anger over the extremely controversial reintroduction of the Aboriginal hunting of grey whales at Neah Bay in Washington State. Important debates about Native Americans' right to sovereignty are sidestepped or converted into debates about ecologically appropriate or inappropriate behaviour.

"Logging Culture Is Like Indian Culture"

A final point in the troubled, competitive, and tense interactivist dialogue about cultural legitimacy is that loggers malign the ecological-nobility thesis at the same time as they cast themselves as analogous to Aboriginal peoples and, therefore, as worthy of respect. The stigma-rooted quest of loggers for grassroots legitimacy was established in Chapter 4. Here, the pursuit of respect extends equally to questions of cultural legitimacy. The right to speak authoritatively about Oregon's land-use practices is based upon not only the stigma that loggers suffer, but also upon alleged parallels drawn between the circumstances of loggers and the circumstances of Aboriginal peoples.

One means for asserting this claim is to protest, as do many, that logging is a culture like "Indian" culture. In the words of Steve Fuller, logging "is a culture and a way of life. It's in the heart, it really is. It's in the heart; people just want to do it. It's as culturally ingrained as Indian culture." Here, cultural legitimacy rests upon the premise that loggers' beliefs and practices have become so internalized that Fuller and those like him no longer act out their vocations consciously. A number of social scientists have documented the practices and worldviews of loggers (Brown 1995; Carroll 1995; Fortmann, Kusel, and Fairfax 1989; James-Duguid 1996; Lee 1994).

But it is a poignant fact of this debate that (1) loggers must contend with popularized notions of Aboriginal peoples as exemplars of an environmentally ideal lifestyle and (2) must contend with a consistently disrespectful public mood towards loggers, be it based upon incipient classism or the growing tendency among North Americans and Europeans to view loggers as the antithesis of a "sacred" approach to nature. Given this social climate, loggers were often compelled to assert their own cultural legitimacy by drawing dubious parallels between themselves and the very Aboriginal peoples they elsewhere seek to discredit.[24]

An example involves loggers' frustration with negotiations between the timber industry, environmentalists, and the federal government regarding the level of timber harvests on public lands. Loggers equate their perceived mistreatment with that experienced by Aboriginal peoples in relation to treaty negotiations in the American west.

Initially, the possibility of turning to public lands for timber supply was enshrined in the late-nineteenth-century acts that enabled the national forest system. Subsequent legislation – namely, the Multiple Use Sustained Yield Act, 1960; the Wilderness Act, 1964; and the subsequent RARE I and II agreements were supposed to have decided the availability of public lands for logging.[25] For this reason, Frank Kramer, part owner of a small old-growth mill nestled in western Oregon's coast range, very nearly kicks himself for his naiveté concerning loggers' repeated negotiations with the federal government.

> We should have seen it coming because it's been, like I was saying before ... [we] hardly got the ink dry [before] we started arguing over roadless areas. Every time we tried to build a road in one of those leased areas, they get an injunction, strap themselves to the trees, blow up the equipment, and fight over them. In the last six years or so, its gotten really more intense, at least in my life. When I got out of forestry school, the whole forest, all the national land here, was available for [pause] land use. [Frank appears to pause for a moment and think about his choice of words, replacing "land use" for what, in less sensitive times, would probably have been "logging."]
>
> They were all dedicated to the multiple-use principle and the sustained-yield principal and all those great things from Teddy Roosevelt on. But then they started setting aside what they used to call wild areas, some of them went into wilderness areas. We've had that kind of an argument ongoing for at least a couple of decades, and it's always "they want more." It enlarges and enlarges, taking more and more. And the last time, we thought we had it solved. In other words it was a compromise, there was *real* compromise. In Oregon it was a million, two or three hundred thousand acres, a huge area, into wilderness. The other areas were going to be managed for timber, and they put this into law.

This experience has led Frank and others to propose defensively that, after years of battle, loggers assumed they were at a "resting place," only to be faced with the much wider restrictions on logging introduced under President Clinton's forest plan. In the scheme of things, the more recent Clinton plan is perceived as the heaviest link in a long chain of negotiations, each of which ended in increased restrictions on logging. It has also fostered the contention that loggers are akin to Aboriginal peoples in that they have been assuaged and misled into non-binding and thus unreliable agreements. Further, loggers see themselves as unequally treated by environmentalists, whom they view as willing to support treaties concluded between the federal government and tribal groups yet as unwilling to support logging "treaties." Beverly Mason puts it thus: "We felt that once we made an agreement that that was it. We didn't realize that the minute you walked away from the table, the environmentalists were going to file a lawsuit." Beverly then goes on to say that "they [environmentalists] can say that they didn't go back on their treaties with the Indians," but they cannot say the same of their promises to loggers.

The history of Aboriginal land rights in the Americas is not equivalent to the history of loggers' "rights" in the Pacific Northwest. Any perusing of the historical record will attest to this. But loggers have been led to believe, by a cash-hungry Congress and overly optimistic federal land managers, that logging would continue unabated forever. The law "promised" federal lands to all people and interests, not solely to loggers and the timber industry, but, through most of the twentieth century, non-timber uses were virtually ignored by the Forest Service (see, especially, Hirt 1994).

The "Right" to a Cultural Past and Future

What do loggers have to gain by critiquing environmentalist efforts to valorize Aboriginal practices or by presenting themselves as analogous to Native Americans inasmuch as they (loggers) have been denied logging "rights"? First, it reinforces their desire to remain visible within the larger cultural and historical account of human habitation in the American west. It helps defeat the portrait of loggers as transient curiosities or unrooted incorrigible cases. Second, it helps to amplify their presence in the long tradition of land management in the region – Aboriginal and not – thereby fulfilling the maxim that "to control the vision of the past is to control the possibilities for the future."

Logger Stu Hardesty spoke directly to these points when, late one winter afternoon, we stood talking outside the rural home he owns with his wife Jean. Earlier that day, I'd been given a tour of their property and spent much time examining the rusting antique logging equipment that Stu collects. Stu likes to point to nearby hillsides and forested plateaus when specifying cardinal directions or describing past events. It was as though he was trying to help me conjure up the small village or westerly situated country store that was the subject of his story. Like so many loggers, he also recalls with excitement his reading of Lewis and Clark's westward journey. What captured his imagination, in particular, was the sheer volume of people who once occupied Oregon Country and his own membership in the ebb and flow of inhabitants. Expressing his delight with a recent publication of the Lewis and Clark expedition journals, Stu offered the following:

> They're really interesting. What they talk about in there are the different Indian tribes and the Indians they met. And what I was surprised about was it seemed like every fork of the river, where another river came into the Missouri, was another nation. Not just a tribe, but a nation. You know, with 10,000 Indians living there. And I think about what happened to all of those people.

Stu also liked to note that the evidence for an abundant past population and numerous prior cultures asserted itself on the very land upon which we stood, manifested in the artifacts that he has uncovered in the area. His vision of the past refers not to decades ago but to something deeper, more enduring. Chronicling local history, he offers a trajectory, from the Aboriginal inhabitants of earlier centuries to the generations of early American pioneers from which he is descended. "People look at something and say holy mackerel, you know, you should have seen it fifty years ago." Dismissing fifty years as comparatively recent, Stu points instead to the burned log that he and his brother uncovered.

> Right up here on this flat above my house – my brother's got an old pound rig that we'd drill wells with during the winter. I went down 110 feet and I couldn't go any further. Hit an old-growth yellow fir log. It was burnt, and I'd bring up the burnt chunks, you know. They [Lewis and Clark] talk about how they burned – you know, they would burn, set the prairies on fire to get the game and the timber.

To Stu, the burned, submerged log, unearthed while digging a well, is symbolic of the Aboriginal practices that preceded his own land-based existence. His further comments address his affiliation with subsequent logging and homesteading populations.

> Well, the old wagon road that used to come from back up there on the hill is out here in my pasture. I asked the old guy I bought this place from, I said, "What is that old road there?" And he said, "That's the old wagon road." He said they weren't using it when he came here in 1930-something,

when he bought this place. But he said, "That road used to go up to – you know where the north fork tower is up there?" I says, "Yeah." And he said, "Well, there used to be a store up there, if you can believe that." I mean there's nothing up there [now]. There's timber growing everywhere. But that was all cut back then.

Stu's historical sketch does not acknowledge the colonial period in the American west during which the Aboriginal populations either succumbed to disease or were removed to reservation lands – an epoch that historian Elliot West refers to as the "greatest die-off" in the area's history. Instead, the overall effect of Stu's multicentury sketch is to cast himself and those like him as members in one long trajectory of human habitation, each following the other in a relatively steady and seamless stream of humanity. Moreover, he establishes that a once-flourishing population – first Aboriginal and then White – has diminished to the point of biotic reclamation. To say "there's timber growing everywhere, but that was all cut back then," is to say that timber has replaced a once human-dominated landscape, further testament to the region's loss of rural land-based communities.

Fire, Like Culture, Is Sui Generis

The view that logging communities are central to, and even seamless descendants of, land-based traditions in rural Oregon is crucially important to loggers not only because it locates them as cultural successors, contemporary claimants in a linked arc of cultural practices, but also because it portrays the arc of which they are a part as sui generis. The arc, they say, has a life of its own and, de facto, should not be preternaturally disrupted. Loggers see themselves as having begun/continued a historical tradition in its own right, and they argue that disrupting that tradition's momentum is akin to showing cultural disrespect and, more important, promises ecological disaster. This point deserves clarification because it is here that much of what has already been asserted by loggers – that Aboriginal peoples managed land aggressively with fire, that loggers are intrinsic to a larger arc of historical practice, and that loggers seek respect as a cultural phenomenon – comes together as part of their greater effort to establish themselves as the architects of future land use in the region.

In Chapter 5, loggers were represented as good land managers, careful and artful farmers of the forest. In this chapter, loggers present themselves as having improved upon nature and prior cultural practices – a claim manifest in their success at recovering once-burned forested areas. Logger Ronald Gautier testifies to this as follows: "Historically, this [area] – the foothills of the west slope of the Cascades – has more timber now, I think, than it ever had as far as ground-growing timber because of the fire history. The Native

Americans used to just burn [inaudible] country up there, you know. The old timers said they can remember it, standing in one spot they could see for miles and now all you see is trees."

For loggers and environmentalists alike, repetitious, mantra-like references to fire point to their mutual efforts to define future land management by insisting that each has the best recipe for repairing or recovering nature. Bettering nature by adding forest cover is a debatable point. But if one's measure for improvement is greater forest acreage – a logical conclusion for loggers, for whom "forest health" means a robust timber-reproducing forest – then Ronald's evidence is compelling. If, however, health is defined as a forest that most resembles naturally occurring cycles of decay and recovery, then a healthy forest is one in which non-anthropogenic fire (e.g., lightning-ignited) has been allowed to burn, periodically reducing the undergrowth that constitutes a fire's fuel load. Regardless, in the American west, fire has been actively suppressed for most of the twentieth century. The legacy of this is an abundant accumulation of combustible material on forest floors. If, after years of fire suppression, new fires are left to burn "naturally," the result could be catastrophic.

This is also the reason that fire acts, for loggers, as a perfect metaphor for culture. Fire, like culture, has its own momentum – neither should be decommissioned. "Mankind," offers logger Jim Stratton, "can only perpetuate what he started. He's got a tiger by the tail, you might say, he can't let go of it, he's not going to let fires burn." The amalgam of ideas about fire and culture is, in turn, articulated by Bill Hawkins (introduced in Chapter 4), with the ultimate goal being to maintain respect for current – albeit changing – "cultures" of land management as defined by loggers and industry foresters.

The Pacific Northwest was never a sea of old growth ... [There are places] where it [old-growth volume] was 30 to 40 percent because of fire and Native American interaction. Knowing as much as I do about ecology, there's no way we could have had a sea of old growth as perceived ... Humans ... have an interactive role ... To just say leave it alone, nature is the only way, I think, does a disservice to everybody who's lived before and given us knowledge, and everybody that's going to live after us. That's a historical philosophical view of mine that I think is different from a lot of those in the environmental community, that's the big difference.

In one linked train of thought, Bill contests the possibility of authentic wilderness ("we never had a sea of old growth") and the valorization of Aboriginal practices (they are partially responsible for reducing the region's volume of old growth); he also condemns as culturally disrespectful the misrecognition of knowledge held by loggers and foresters.

〜

In Chapter 4 it was obvious that forest-community activists were trying to gain purchase in the forest debate by establishing themselves as legitimate grassroots players; further, the need to do just that indicated some uncertainty regarding their status as bona fide activists. Ancient forest activists appeared comparatively confident about their grassroots status but did reveal some unease in the face of criticisms that they were indifferent to the plight of loggers. In Chapter 5, two perspectives on science were revealed, each of which suggested very different approaches to the "management" of public forests. In this chapter, I demonstrate that both loggers and environmentalists recognize the symbolic resonance, and thus discursive power, of references to Aboriginal land-use traditions. This (symbolic) power is rooted in mainstream American notions that Aboriginal peoples represent the possibility that there are peoples who effortlessly embody principles of Western conservation (Conklin 1997, 722).

Even though many students of culture have come to regard the nobility thesis as superficial and/or as variably appropriate, stereotypic references to it have not vanished from the activist lexicon; rather, activists from both groups recognize that the legitimacy they can claim regarding past and future land use and/or the ease with which they can play into mainstream notions about Aboriginal peoples affects their rhetorical leverage and their capacity to imagine cultural change. Practically speaking, this means that those most closely affiliated with the Aboriginal, or authentic, tradition wield a distinct political advantage.

Yet the question of legitimacy presents a problem for both activist parties because both have markedly equivocal claims to place-based authenticity. Environmentalists are disadvantaged on this point because only a very few have hands-on/lived-in experiences with the forests they seek to protect. Very few live outside urban areas, few extract a living through physical labour in the natural world, and most are several generations removed from the Euro-American pioneers who "replaced" northwestern Oregon's Chinook and Kalapoia tribes. One way of gaining an advantage under these circumstances is to ally one's group with those who already possess authenticity, thus vicariously acquiring a degree of cultural capital. Discussions about past peoples, like discussions about grassroots legitimacy, can be used to bolster one's own fledgling cultural authority through the sheer force of being associated with more respected or more "cultured" groups, especially those that are believed to emulate desirable options for the future.[26] And yet, activism is not simply about language and discursive advantage; it speaks equally to real behavioural change. Hence environmentalists do offer self-criticisms of their own stereotypic practices while simultaneously seeking viable land-use models that might eventually supplant aggressive timber-harvest policies and modernist tree-farming landscapes. That search is often inspired not only by science's insights into the complexity and mystery of forest ecosystems, but also by an affinity for deep traditions and the knowledge born of long-time dwelling in place.

Loggers, conversely, appear uneasy with the strategic advantage embodied in the ability of environmentalists to capitalize on mainstream American notions of ecological nobility and to benefit from political affiliations with Aboriginal activists. When addressing the behaviour of their opponents, loggers seek to redefine popular images of an "Aboriginal" past while simultaneously recasting themselves as analogous to Aboriginal people and, thus, as deserving of "authentic" status along with any rights that accrue. Loggers have their own vision of future land use, which they regard as rooted in a more recent past – one that improved upon Aboriginal practices. They regard attempts to mimic Aboriginal traditions as introducing (not alleviating) ecological danger, and they equate the introduction of Aboriginal-inspired ideas with a naive and non-natural disruption of the practices and communities of labourers that carry out those practices.

7
Irrational Actors: Emotion, Ethics, and the Ecocentred Self

In each of the previous chapters tentative hints of the gasps and pulses of emotion have surfaced, from the affective cast characteristic of the stigma experience, to the juxtaposition of emotion with reason (in the discussion of science in Chapter 5), to several vitriolic comments from activists on both sides. Moreover, at least two of the incidents that took place at a protest encampment (see Chapter 3) were notable for their affective resonance; namely, the uneasy embrace of the old-growth tree that two others and I participated in, and the tense exchange between a logger and two male environmental activists. The second incident serves as a lead-in to this chapter's discussion of emotions, ethics, and the ecocentric self. To recapitulate: A logger disembarks from his four-wheel-drive pickup truck to talk to two environmentalists who are standing near their parked cars. Their three-way conversation becomes tense as the younger of the two environmentalists insists that he understands the plight of his "bro," the logger. The older environmentalist senses that such statements of camaraderie will exacerbate the growing tension between loggers and environmentalists and suggests that he and the logger walk through the proposed logging site to discuss its biological merits.

Those with even a passing familiarity with 1990s Oregon will recognize that variations of this confrontational scene played out repeatedly on television news programs throughout the period. In some mediated instances, logging crews would be hovering at the base of a giant cedar, fir, or sequoia, awaiting the descent of an activist tree-climber who'd settled him- or herself high in a tree fort, hoping to save the old-growth giant from its mill bound destiny. Equally popular with the media were dramatic logger-environmentalist showdowns, which invariably confronted viewers with graphic images of environmentalists chained to a Forest Service gate or road-building equipment while they were berated by a logger for interfering with his work. Or the scene might involve two antagonists (often both male) exchanging invectives, only to have the logger (identifiable by his yellow hard hat and heavy boot safety gear) portrayed as punching or shoving his opponent. Generally speaking, these and other images fed the non-activist public's belief that the controversy had become too emotional, that activists had lost their "rational heads." Loggers were increasingly viewed as unable to contain their rage, while environmentalists perched in trees or embracing a tree's trunk were viewed as having become too "woo woo" (a popular local phrase used to characterize the affective and bodily loving of nature).

This penultimate chapter takes emotion as its central subject. This is necessary both because emotions are central to understanding collective action aimed at changing some aspect of society and because emotion, as a field of study, has largely been neglected by students of social movements (Jasper 1998). I begin with the assumption that the study of emotion offers a unique lens through which one can view everyday moral discourse about what constitutes appropriate human behaviour in the social and natural world. Emotions generally accompany expressions of judgment and proposed action about how one imagines both the present and the future world, and/or how one believes the self and others should act in that world. This raises two questions: (1) What clusters of convictions and promoted ethical practices accompany the forest dispute's emotional content? and (2) How does an examination of emotion further illuminate activist positions?

On the Social Study of Emotion

Emotion, as a subject worthy of philosophic, psychological, and biological investigation, has a lengthy academic history, reaching in Western intellectual traditions at least as far back as Aristotle. As the antithesis of logic and reason, emotion has served as a useful counterpoint against which more desirable reason-endowed categories of behaviour could be judged and studied. Also, particular to this view is the belief that emotion is a purely physical and psychic phenomenon, a product of the internal world of every human being. The corporeal lurches, racing heartbeats, and generally overwhelming nature of some feeling states are said to signify the human body as "wired" to fear (and flight), anger, lust, and a host of other passions. On these grounds, scientists have theorized emotions as precultural, individualistic, idiosyncratic, and bodily rooted – as interesting and perhaps inconvenient eruptions in psychic and social life but otherwise beyond the scope of social analysis.

In the last two decades, however, this position has been substantially revised as anthropologists, sociologists, and social psychologists have begun examining the many ways in which emotion can also be said to be socially and culturally constructed (Planalp 1999; Shweder and Levin 1984). Elaborating upon what this claim means introduces one to a rich literature, which I will touch on briefly here and throughout this chapter.[1] To say that emotion is socially "constructed" is to say that our emotional behaviour, the quality of our emotional demeanour, is strongly affected by social factors. The human body is the site of emotional expression and activity, but emotions do not get expressed in a fashion that is idiosyncratic to the person emoting; rather, it is more accurate to conceive of emotions as managed, consciously or not, by the individual agent who uses a style that closely reflects identity- and culture-specific codes for appropriate emotional behaviour.[2]

Just what rule of emotion applies to whom and under what circumstances remains an active area of research. Among the earlier and more influential findings is the contention that the rules for feeling are based upon one's structural position in society; that is, upon one's class, gender, caste, race, age, and/or ethnicity. In a set of studies on the emotional comportment of airline hosts and college students, Hochschild (1979, 1983), for instance, posited that, in many circumstances, we learn to deep act so as to avoid displaying what we "actually" feel in order to abide by situationally appropriate behavioural rules. She found that White, middle-class women effectively sell their emotional labour to airline companies because they have been socialized (and can be further trained) to contain their anger and to remain gracious in the face of belligerent customers.

Other emotion scholars have gone on to highlight the fact that with social power comes the ability to act out emotionally with minimal repercussion. Within mainstream American contexts, people in positions of power or status tend to be granted, behaviourally speaking, greater emotional latitude, whereas people with less status are expected to be more conscientious and/or more pressed upon to restrain themselves emotionally (Lutz 1986; Planalp 1999). "The metaphor of control implies something that would otherwise be out of control, something wild and unruly, a threat to order" (Lutz 1990, 72). The maintenance of emotion-imprinted power relationships is served, moreover, by castigating marginal persons as being emotionally out of control, thus justifying the "discounting" of their concerns. Hence, the poor are dismissed as crazy, whereas the rich are merely eccentric; activists outside formal offices of power are irrational, while elected officials with like views are thoughtful and challenging.

Loggers and Emotional Control

Within the Oregon context, and for activists everywhere, the possibility that they may be dismissed as unworthy of consideration on the basis of emotional comportment is a salient one. Generally speaking, it translates into a heightened awareness and consideration of how to conduct oneself emotionally vis-à-vis the forest dispute. For loggers, this was most evident in their frequent talk of the need to control their anger so as to withstand the view of themselves as socially dangerous and politically irrelevant – a view derived from their working-class status and stigmatized reputation.

Routinely, in fieldwork and interview contexts, loggers emphasized the benefits of a reserved emotional climate. So it was that small woodlot owner and logger, Ronald Gautier, praised President Clinton's Forest Summit for its capacity to demand an atmosphere of affective restraint, to demand that participants' "best" (i.e., most "rational") behaviour prevailed. "They didn't," observed Gautier, "just sit and scream at each other."

A more detailed interactive example is offered by Arlene Baker (whom we met in the previous chapter), her son Chris, and her husband Gary. Their unedited comments were elicited when I asked them what they thought about the timber dispute's emotional content and asked them to consider whether that content was a good or bad thing. This afforded them an opportunity to self-direct their speculations about this feature of political life and/or to project onto my open-ended question their own evaluations, experiences, meanings, or ethno-theories about the dispute's affective qualities. In the following passage, notice in particular the Bakers' poignant effort to deflect accusation of excessive emotionality and/or to assign it to their opponents.

(01) *Gary:* I think there's a lot of emotion but it doesn't really do any good. It don't settle the problem.
Arlene: It never has and it never will.
Gary: That's not what gets the job done anyway.

(05) *Chris:* No, but the – well, people get pretty – the other side gets emotional and I ...
Gary: ... Real one-sided.
Chris: ... call them the misguided people that I call them. There's a lot of young people involved that have bought into all the rhetoric and think that if

(10) they don't do something, you know, disaster's coming, and therefore it's a very personal thing to them, you know. And it's very personal to us, you know. When the paycheque stops flowing you get pretty upset. You know, people get pretty violent even, you know. Because they see ... [overlaps with line 16]

(15) *Arlene:* You have to be a bit careful.
Chris: ... Matthew Waite on the TV and they – or if they go to a rally, you know, like we went to rallies and stuff, and most guys who have ...
Arlene: Kept themselves very well under control.

(20) *Chris:* ... kept themselves under control very well. But I'll tell you, it's hard for these guys to get face-to-face with the guys who have stopped their paycheque and not seriously want to hurt them.
Gary: ... Not only their paycheque, but their way of life. You know, this, we've lived like this for years.

(25) *Arlene:* Yeah. That's the emotional part for me, is to watch people, older men who are having to take orders from some young smart punk.

And you stand and you watch and the times that I have become emotional was like the time that I thought that I was going to go over –
you guys were arguing with that little weasel in front of the Wilderness Society [booth] – and then I

(30) became emotional when he called me a name. So, and you can lose it
really easy. But then you have to kind of pull yourself back in and say
no, I'm not going to buy into this.

Emotion, in this passage, is first synonymous with human conflict. Equally
evident is the conventional claim that emotion interferes with action (or
"getting the job done"). But it is the persistent emphasis on emotional control that dominates the passage, a claim that echoes allusions in earlier chapters to loggers' reputations as "drunken, brutish, louts." Having introduced
emotion as a counterproductive ("job-interfering") force, Chris moves (in
line 05) to maintain his emotional acumen by reworking his first impulse
to include himself and his cohorts among those who "get pretty ... emotional" and, instead, to define this as a problem exclusive to the "other
side." The human subjects in his narration change from "people" in general
to "the other side," which allows father and son to present themselves and
their allies as rational and controlled. It is a conversation that says: "We are
the ones willing to negotiate; we are fair and sensible players in this game."
   Chris then proceeds to describe his opponents as having succumbed to
emotion (i.e., "disaster rhetoric") and, by definition, less honorable feelings. On the occasions where "inappropriate" bursts of emotion slipped
into the Bakers' conversation (e.g., lines 19-30), it became necessary for
Arlene to take responsibility for the outbursts. Widespread, and admittedly
essentialist, assumptions about women as appropriate agents of emotion, as
"more emotional than men," enables Arlene to become emotional in a protective and then violent sense when faced with the insult of having the
men in her life denied work (let alone "emasculated" by someone young
and ideologically abhorrent to her). She uses her gender (and the fact that
she is not a logger) to resist pejorative stereotypes about the wanton anger
of loggers while simultaneously portraying protective (traditionally "female")
instincts by upholding their masculinity ("little weasel" versus "older men").

Agency and Emotional Control
It would be wrong, however, to suggest that activist behaviour and affective
demeanour can be neatly mapped onto the sociostructural (especially class-
based) positions they correspond to most closely. There are many excep-
tions to this rule, which, of course, force me and those interested in emotion
to consider the possibilities (and theoretical implications) of behaviour that
resists the rules for emotional management associated with a particular group.
For instance, Reddy (1999) offers examples of emotional behaviours

standing in tension with, or contradicting, prevailing norms. Many of his examples are drawn from societies with strict codes of behaviour, such as those characteristic of authoritarian social structures. He reminds us that, in order to succeed, every emotional regime must allow for wide personal variation. The subject of variation raises two linked points important to both emotion and identity theory, and renders pertinent a synthesis of the two.

The first point is that core cultural themes – such as that of emotion or any of the other master cultural themes taken up in earlier chapters (democracy as it pertains to grassroots legitimacy, science, cultural authenticity, etc.) – usually embody several possible meanings. Emotion can be defined in a number of different ways and, thus, offers social agents a certain latitude when deferring to or invoking affect. For instance, emotional control is an important social norm in North American society. We value the possibility of a political arena characterized by cool reason and civility. But, like the psychologist William James (1901), we also cannot imagine and so do not wish to live in a world of Spock-like passionless automatons.[3] Emotion, in this sense, is framed by two poles of expression, with positively construed control at one end and sometimes positively construed passion at the other. Feminist emotion scholars have been particularly good at pointing out this dichotomous structure in much of our thinking about emotion (Lutz 1988; Strathern 1980).[4] We tend, they note, to organize the world into positively versus negatively construed opposites so that emotion, women, and nature are never quite as important or valued as are reason, men, and civilization. Women, unlike men, are often viewed as lacking rational thought. Yet these same women may be redeemed when emotion is contrasted with cold alienation. "Emotion, in this view, is life to its absence's death, is interpersonal connection or relationship to an unemotional estrangement, is a glorified and free nature to a shackling civilization" (Lutz 1990).

The second point, following recent identity and agency theory, is that, in all societies, behaviour (in this case, emotional behaviour) can be construed and carried out in a number of ways, albeit in a culturally and structurally limited number of ways. A group or society's culture (including its rules for emotional conduct) is not uniformly imprinted on all its members; rather, individual agents encounter the rules in situationally unique circumstances and, thereafter, author the specifics of just how to take up the cultural prescriptions within which they operate.[5] The identity movement with which the agents are affiliated, in turn, informs this taking up. Identifying as a logger or an environmentalist, for example, figures prominently in how the structural (race, gender, class, etc.) rules affiliated with a core construct such as emotion are realized. Practically, this means that discourses on emotion "become the media around which socially and historically conditioned persons" conduct themselves (Holland et al. 1998, 32). It is a creative, though not necessarily conscious, process that reflects the salience of social forms (e.g., the salience of rules about class and emotion) across time and place as well as the situational richness that characterizes human interaction.

Activists' creative, though structurally limited, adoptions of the normative rules for emotional behaviour can be illustrated first by examining loggers' counterpoint musings about emotional control, and then, by focusing on the behaviours and meanings that environmentalists attributed to emotion.

Loggers and the Wisdom of Emotional Obedience

After the Bakers' first response to my query about the role of emotion in the forest dispute – which involved an unabashed focus on affective restraint – I found them actively questioning the wisdom of their emotional "obedience." In the passage that follows, one can observe the back-and-forth swing that leaves them uncertainly suspended between being emotional and yet not emotional "enough." Cognizant of their marginality and, thus, their vulnerability to further denigration should they fail to "remain calm," the Bakers also worry that loggers' practised restraint (and the resultant view that they are an emotionally estranged people) has kept them from promoting what they regard as significant group qualities: (1) masculinity and (2) the depth of their collective despair. An emphasis on emotional restraint, the Bakers find, causes loggers to come across as "not hurt that bad," while environmentalists, whom they characterize as emotionally excessive, "become heroes to some." This engenders a certain ambivalence for loggers (about remaining calm) that the Bakers must somehow negotiate – an ambivalence that captures very neatly the contradictory co-valorization of reason and passion. Just how they negotiate this draws attention to their agency and, hence, to their ability to work within the window of opportunity that contradictory rules and definitions of emotion provide.

A first example is provided by Arlene, who suggests that loggers' self-control and/or suppression of aggressive emotion has served only to downplay their "superior" masculinity. (Better this be done by Arlene than her husband or son because, as a woman, she can more readily "get away with" abandoning the necessity of emotional control.)

> Now see I think what bothers me in a way – and maybe we should have handled it from the beginning this way ... I'm not sure [but] maybe we shouldn't have in the way that we've just been so laid back and try to be so nice ... I remember when I was really little and the old time loggers and my dad was – why they wouldn't have put up with it for a minute. They'd have handled this themselves. They'd have just went in and whupped the tar out of them, and every time they came at them there would have been a battle and that would have been it.

Chris is, alternatively, more cautious. He wishes to abide by the dictum that reason will serve his cause, yet wonders aloud whether environmentalists haven't capitalized on "emotion" rather than suffered for its use. Speculation along these lines were first suggested when Chris earlier disparaged his opponents for their "disaster rhetoric." In the lines that followed (10-11), he granted environmentalists a modicum of integrity (they "take it [the rhetoric] personally") only to have to then quickly attribute a comparable degree of emotional integrity to loggers ("it's very personal to us [too], you know").[6] Later in our conversation, he returns to this point explicitly.

> The only thing about it is they – I sometimes think that they've got the – this issue by being emotional. You know, by sitting in the road in front of logging trucks and chaining themselves to trees and what not. And sometimes I think, you know, *geez, maybe we should get more emotional,* because they've – they're the – you know, they've, by chaining themselves to the trees and, you know, just burying themselves in the middle of a logging road and *blowing up bridges and stuff.* They've almost become heroes to some, so why haven't we ... Sometimes I wonder if our lack of emotion sometimes makes it seem like we're not hurt that bad.

Chris's speculations are part question and part accusation. They also point to the juncture where emotional norms and agency meet. When he asks why "our side" is obeying social norms that favour reason over emotion, one can hear the influence of normative rules that prescribe rational behaviour. When he follows with "they've become heroes to some, so why haven't we," human agency surfaces in the form of social critique, protesting that the rules involve an unfair double standard and should be changed.

The tension in Chris's commentary between "remaining calm" and the possible benefits of extreme affect is an example of the co-valorization of reason and passion – a contradiction that sets up an opportunity for the Bakers to engage in an act of "authoring." This is an example of the windows of opportunity, the interface between agency and culture, that identity theorists Holland et al. (1998) have termed "the spaces of authoring." The parallel theoretical point, as it applies to emotion, is as follows: Chris is operating in a cultural world that generally expects people to behave rationally in civic life but that asserts that prescription to a greater or lesser degree, depending upon the marginality of the agent's position and the multiple meanings (e.g., indifference versus engaged) ascribed to emotion. We already know that loggers, due to their class and local stigmatization, occupy a distinctly marginal position. Chris was, for this reason, significantly confined by his social position, especially on the point of emotionality. Nonetheless, he manages to strategically author a critical point or two about how loggers suffer for their obedience relative to their competitors.

He works within the confines of his position to achieve sympathy and to critique a system that allows his environmentalist opponents more emotional latitude than is granted to loggers.

Relational Emotions

Environmentalists, like loggers, were astute agents, equally able to take advantage of the class- and gender-specific rules for emotional behaviour. Their (emotional) agency took the following forms: first, environmentalists were comparatively disinterested in the restraint of anger – an indifference that reflected their comparative advantage over loggers. By "advantage," I refer to their social power as non-stigmatized, predominantly college-educated, middle-class activists. When I asked environmentalists to discuss emotion generally, they turned instead to comments about emotive relationships to nature. They assumed that my asking about "the dispute's emotional content" meant that I was asking about their affective and spiritual attachments to nature. It was also apparent that environmentalists embraced rather than resisted emotion, particularly if female. Monica Ladner, for instance, behaved as though my open-ended query about emotion referred precisely to her belief that to be "emotional" within the context of the forest dispute was to be "spiritual."

> I think that the emotional parts of [the dispute] are the closest that we've sort of come to, uhm, a spiritual place for doing the work, as far as the movement, kind of in general. And that there remains sort of cultural taboos in our society about being too vocal on that. So there's still a huge concern about being at least vocally or visually connected to that ... But a lot of, you know, the majority of the activists that I know, leaders of the environmental movement, do spend quite a bit of time out in the woods, you know. Camping or hiking or, you know, whatever it is that they do. Including hunting, some of them, so that they do, on a personal level, maintain some connection with [nature].

Ladner's argument is that the "emotional parts" come about because spiritual motivations for environmental work are prohibited. Emotional expressions are more acceptable than spiritual ones and, therefore, more visible. Her comments also reveal her frustration with the public's (and environmentalists') failure to be more overt, hence honest, about their emotive-spiritual orientations towards nature. Also implicit in her comments is the notion that if one is not cultivating an emotional attachment to nature, then one ought to be. Finally, the coupling of forest recreation and emotional "connectedness" in her last two sentences echoes historians' observations that environmentalist discourse is imprinted with romantic traditions in which forests, particularly those designated as wilderness, are construed as divinely endowed and are regarded as a key source of spiritual inspiration: "the rare places on earth where one ... [might] glimpse the face of God" (Cronon 1996, 73). More recently, they are seen as places where one might escape the debilitating effects of an industrialized and urbanized civilization.

For Ladner, being in nature has come to mean recreating or restoring one's spirit at the exclusion of interacting with nature for material benefit or need. It is a point of considerable annoyance for most loggers, a resentment that manifests itself in their perceptions of environmentalists as morally arrogant. Monica Ladner, in fact, appears to consider the ire of loggers when she acknowledges, in the passage below, that timber workers possess their own "feelings for nature." But their capacity for feeling is then largely negated by her subsequent claim that loggers' utilitarian and "male" perspectives override it.

> And I think it applies to the timber industry too, that, you know, they're not totally isolated from their feelings of connection to the Earth and to trees and whatever. It's just that their livelihood has become dependent on living in a way that is a result of a cultural way we look at livelihood and look at, particularly men's identity, that is tied up with producing, with having a job, with controlling, dominating, whatever. And that even though they feel that connection, the overriding identity is around their male identity.

When Ladner attributes the domination of nature with production-oriented practices and the "male identity" generally, she reflects the basic premise of ecofeminism: the domination of nature is intricately linked to the domination of women by men (Merchant 1992; Warren 1988, 1990). A feminist ethic posits that much inequity can be reduced by advancing women's rights, defined broadly as improved access to and participation in all spheres of public life. An ecofeminist politics suggests, alternatively, that a rights-based program should be augmented by an ethic of care, or nurture, between all humans and between the human and non-human worlds. It is a female-centric ethic to the extent that women are regarded as more capable of caring behaviours than are men; their reproductive roles and/or experiences as mothers are believed to prepare then more fully for becoming "nurturers" of nature.

The proposition that women are closer to nature is, like the ecological nobility premise discussed in Chapter 6, an essentialist one, and it incorporates the many pitfalls that such assumptions embody.[7] Nonetheless, the idea is widely promoted by both women and men in Oregon's ancient-forest movement. Indeed, all but one of the female environmentalists consulted for this study emphasized the need for an emotive-spiritual relationship to nature in order to offset environmental decline.[8] The practice is both ideologically compelling and strategically essentialist in the sense that a widely held stereotype is reworked by the heretofore "victim" of such stereotyping so as to positively empower her within the environmental movement.[9] Moreover, as with loggers above, women are authoring their use of core cultural concepts about women/nature and emotion to beneficial ends. By amplifying the significance of women's experiences in a characteristically male domain (e.g., the environmental movement), they convert the traditionally devalued position of "females" to a valued one.

That some "feminization" of the movement had indeed taken place was apparent in that slightly more than half of the men I interviewed spoke of cultivating their emotional and, by implication, female sides and/or spoke favourably of an emotive-spiritual relationship between humans and nature. But, as men, the position-specific authoring of their emotionality differed. Michael Costas, the logger-turned-environmentalist whom I introduced in earlier chapters, believed, for instance, that an emotive-spiritual relationship with nature would offer a solution to what he deemed to be "society's existential crisis."

> One of the things that is very important to me is that I look at human beings right now in this society that we've developed, I am incredibly dismayed by how much, how many people are walking around with spiritual and emotional holes inside, that are trying to fill them up. And they're filling them up with drugs and alcohol, and consumer products, and, I mean, it's classic. If you feel depressed, you go to the mall and go shopping ... [any relief is] gone by the time you unload the car when you get home, you know ... But if you go into a forest ... and you sit down and you enjoy it and just look down, look up and just feel it, that hole goes away – it's not there, and it will sustain itself after you walk away from it.

Michael's point is an important one as it draws attention to the way in which consumerism brings to life (or, to use Taussig's term, fetishizes) inert commodities by falsely treating consumable items as though "they were alive with their own anonymous powers" (Taussig 1980, 36) and can thus act as therapeutic agents. Alternatively, for Michael and many others (myself included) the forest is the mall's antithesis, the newly established place where one can be restored by "feeling" its presence. To feel suggests that the forests (rather than consumer goods) are themselves coming to life in the mind of the perceiver and that Michael and others have come to sense more fully that which the forest projects towards the (human) perceiver (Abram 1996).

With regard to feeling, several male environmentalists granted excessive emotionality special status as a means of ridding oneself of the capacity to be indifferent (i.e., unfeeling) towards the non-human world. Positive conceptions of emotions were embraced and embodied in an effort to symbolically and physically counter one's estrangement from nature. A good example was offered at an activist conference during which a nationally recognized environmentalist, Donald Richardson, rallied the attending crowd.[10] In particular, he proclaimed his willingness to forego the political credibility that accompanies emotional self-control:

> Now what I'm talking about is something difficult in modern American society. I'm being very uncool, as my teenage nephew tells me, because I'm being emotional. And you know when I went back to Washington DC ten years ago to work for a national ENGO, a United States senator took me aside and said: "You know, Don, I think you can work out well in Washington. I think you can learn how to compromise and make a deal." And he told me to never be emotional. He told me to put my heart in a safe deposit box and replace my brain with a pocket calculator. He told me to quote only economists and engineers. He said if I was ever emotional I'd lose my credibility. *But dammit I am emotional!!!* [yells] I'm not some new age automaton! I don't have silicon chips up in here [touches his head]! I'm an animal and I'm proud of it.

Yet, towards the talk's end (see below), Richardson's retreat from excessive emotionality was palpable. His indulgent calls for passion and emotional attachment were followed by the hint that such attachments are shameful and that one must temper one's "affective abnormality" with rationality, particularly as it applies to science. Richardson's radical reference to himself as being an animal and proud of it – a reference I read as the emotional embodiment of the natural world – is moderated by his call for the rational actualization of that passion ("the need to apply the emotional side of our brain rationally to ... science"):

> We have got to somehow form an emotional attachment with this Earth ... we've got to be emotional, we've got to be passionate. We can't be ashamed about that, our love for the Earth ... And while we're motivated by that passion and that love, it's time that we begin to apply this side of our brain rationally ... We have got to begin to apply science, we have got to apply rationality, with our emotional, passionate, heartfelt love of the Earth.

A similar pairing of emotion and reason was invoked by other male environmentalists when speculating about the role of emotion in the forest dispute. Greg Norman, who, in Chapter 5, mused about tectonic plates, spoke of his "emotional commitment" to the environmental movement and the concomitant sense of meaning and aliveness he found in that commitment. He summarized his approach to emotion as follows: "If you're really lucky in your life, then you get to employ reason in the satisfaction of your feelings, and you get to employ feeling in the use of your rational faculties."

It is telling that no equivalent effort to pair or seek balance between emotion and reason was evident among female environmentalists. Thus, while positive (and essentialist) conceptions of gender and emotion may be responsible for a growing appreciation of "traditionally female" qualities in Oregon's ancient-forest movement, the emotional rebellion of men is restricted to those aspects of emotion that are positively construed. It is a little like having one's cake and eating it too, in that male activists tie their emotional indulgences to notions of emotional liberation. They aim to free themselves from the estranged human-nature interactions associated with flat affect, yet they avoid being denigrated as irrational by pairing invocations of emotion with an allegiance to rationality and science.

Emotion and Moral Vision

At this juncture it is helpful to step back for a moment, for thus far I have conveyed two impressions. The first is that every social group has its own rules for emotional behaviour and that members of that group will generally acquiesce to those rules. Specific to this impression is the theory that members of a culture are notably confined in the face of rules for emotional conduct. They behave more or less as they should and are quickly brought into line by moral-emotional regimes when they stray from the behavioural fold (Reddy 1999). It is for this reason that we find loggers noting their behaviour and realigning it so as to conform to class-influenced normative expectations that they "contain" their rage. Environmentalists, by virtue of their middle-class status, are less pressured and thus less concerned with practising emotional restraint. They do, however, conform to structurally based norms concerning gender-specific rules for emotional behaviour. Men's discourse on emotions appeals to emotionality *and* rationality, suggesting that they are less willing than female activists to fully embrace emotive relationships with nature. The second impression is that activist parties on both sides of the dispute are constantly challenging conventional wisdom, working within the confines of prescribed cultural norms pertaining to emotion. As active and identity-authoring agents, they are able to work existing structural and cultural prescriptions to some advantage. So it is that loggers are found rethinking the wisdom of remaining calm, while environmentalists are found cultivating a nature-centric emotional life.

These two impressions are consistent with constructivist and self-authoring theories of emotion. The former posits that how we behave emotionally is heavily prescribed by systemic power and cultural norms for conduct. Behaviourally, from a constructivist point of view, the options are polarized: we either comply or resist. Self-authoring theories emphasize that within any dominant cultural system lies considerable room for the self- and identity-specific authoring of conduct. Identity theory, in particular, allows us to see how emotion talk resituates assigned class or gendered social positions. But neither point fully addresses why emotion talk is so prevalent in social movements and/or how that talk operates as a liberatory (following Jasper 1998; Reddy 1999) or change-producing medium. The question is an important one, as social movement theory has generally ignored the centrality of emotion to collective action aimed at changing some aspect of society. Strong emotions are endemic to protest; their neglect as a field of study is thus unjustifiable, particularly as there is a substantial literature upon which to draw for insight (Jasper 1998).

Two points are crucial, and both address the relationship between emotion and moral vision. The first point stems from Jasper (1997, 1998), who claims that activist disputes rely heavily on emotional language to promote specific (and sometimes new) moral possibilities. Emotions arise, he argues, from perceived infractions of moral rules and, more important, activist movements "aim to change the broader [moral] culture ... including the acceptability and display of certain emotions" (1998, 40). The second point is inspired by study of emotion as the central communicative medium through which the rules for moral and ethical practice – particularly about how the human self should behave towards the rest of society – are conveyed (Averill 1994; Lutz 1988; Planalp 1999; Rosaldo 1980, 1984; Stocker and Hegeman 1996; White 1990, 1992; Wierzbicka 1994). Describing to another the emotional experience of losing a job or giving birth is a way of conveying an intimate subjective perspective, but such communications can also assert how one (the human self) *ought* to behave given a job loss or major life event. More fully, emotion talk is cognitively affiliated with scenes that imply morally appropriate behaviour (Lutz 1988). To say, "I am angry about a certain person's behaviour" is to convey a scene in which one mode of behaviour is promoted (the implied desired behaviour) and the other dismissed (the behaviour that produced anger).

Given these stated points, we should be able to examine more closely the change-inducing operation of emotional language and the moral or ethical selves to which it is tied. To do so, consider first the role of specific emotion terms. Lutz (1988) argues for the careful analysis of common meaning-specific invocations of emotion terms. Among her examples are Ifaluk (a Micronesian community in the Caroline Islands) utterances of "song," or "justifiable anger." One night, some men from a neighbouring village frightened island residents with loud and drunken boisterousness. The following day there was much local talk of the event, including frequent references to song. Local references to song were focused primarily upon the expectation that the chiefs (whom Lutz refers to as the final moral arbitrators on the island) will be "justifiably angry" because the offending parties violated an important island taboo. They egregiously disrupted a social life wherein "calm, quiet talk," a "peaceable style," and intervillage harmony are very much the expected norms (Lutz 1988, 157-59).

Specifically, Lutz found that song "scenes" involve a rule violation, the attention drawn to it, a call for condemnation of the act, and the subsequent making of amends. The "prototypical scenario, involves a consistent 'chain of reason' (White 1992, 18).

Social event
(e.g., drunken boisterousness)
↓
Emotion
(song)
↓
Action response
(chief-assigned penalty)

In responding emotionally to an event, here and elsewhere, people are staking a claim as to how things are, and they act to reassert that reality or socio-moral position. The scenario involves an inferential path "that makes emotion talk a moral idiom ... a mode of evaluating and constituting social reality" (White 1990, 48). In this light, emotion talk can be construed as a central vehicle through which social groups continually construct and reconstruct the moral code by which participant members are expected to abide. Emotion talk is the day-to-day micromanagement that continually reasserts and maintains the larger cultural structure (including the chiefs' authority and right to be song/angry, and the importance of song to appropriate behaviour exhibited by outer-village members).

Following constructionist theories of the period, Lutz was principally concerned with the reproduction, through expressions of emotion, of Ifaluk power structures, and so she focused upon the consistency of meaning across utterances of terms. But if we are concerned with agency, and Jasper's conclusion that emotions promote new moral visions, then we should be able to use Lutz's ideas to track activists' redefinitions of emotional meaning (and the moral positions to which emotional language is attached). I expected to find, and indeed did find, conflicting meanings behind emotion terms used by opponents. Loggers and environmentalists may reside in a shared linguistic community (e.g., English-speaking), but their cultural tug-of-war over how one should live with and use Oregon's forests was persistently evident in the moral visions that their emotion-centric language conveyed. These opposing visions demonstrate not only that loggers and environmentalists constitute distinct identities, but also that both activist parties are working to conjure into being disparate moral rules and ethical conduct pertinent to human-nature interactions.

Fear and Moral Meaning: Environmentalists

This is perhaps best illustrated by protagonists' references to fear. Fear has frequently been articulated as "one of the simplest and most natural" emotions because of its association with physical danger (Levy and Rosaldo 1983). When I ask environmentalist Matthew Waite to explain the old-growth controversy's "emotional charge," his response is uncharacteristically brief: "Fear of change, fear of change." The implied accusation in Waite's pithy summary is that fear and, therefore, lack of integrity stands between environmental protection and the change that Waite desires. Fear is invoked as a barrier to action.

A similar point is made by Gordon Hill, one of the movement's more outspoken proponents. Hill's small AFGA-affiliated organization places a degree of pressure on industry and government officials that far exceeds its membership and budget. Saving the environment is a position that Hill, an ex-businessman, believes he can "package, and market" like a product. He also actively distances himself from his more emotionally or spiritually driven peers: "I'm not a warm, fuzzy, lovey-wuvy, love-the-environment kind of guy." Nonetheless, he speaks in emphatic, impassioned gasps. His arms flail outward in a bombastic gesture as he defines himself as "abhorring" compromise. Despite his fervent verbal style, the only explicit reference to an emotion in any of my conversations with him involved utterances of the term "fear."

These appeared in Hill's frequent use of analogy to illustrate the parallels between his experiences in Vietnam – namely, his perception of the political and industrial corruption behind that event – and the political dishonesty that he sees as rampant in the Forest Service and the timber industry it serves. While formulating one such analogy Hill staked a moral claim about fear, the self, and social change.

> I don't think any American can ever conceive of the scope of the corruption in Vietnam. Some powerful, wealthy people became a thousand times more powerful and wealthy, and 55,000 Americans died, and two to five million Vietnamese and Southeast Asians died ... *If there was a God he would have been destroyed over that war* [yells]. Anyhow, I came back and I got combat pay [as a pilot], GI's made $300 a month, I was making like $4,000 a month [even though] I didn't have to suffer combat. I was so chicken-hearted. When I got there a kid in front of me had his head blown off by a stray round, and I had my face splattered with bone and personal effects, bone fragments, and I'm not a brave person in the first place, and went down to ninety-five pounds. I didn't sleep well or eat well. I could eat but I couldn't keep it down ... [Eventually] I was no longer afraid. You know you're going to die, you're going to die. I'd had so much fear, it was just like you are overloaded on fear, and fear no longer played a part of your life. Then I started seeing Vietnamese as people. And it was just horrible, because there

was no demarcation. You couldn't see a good Vietnamese from a potentially threatening Vietnamese. They all wore black pajamas, they all looked the same. They were more cultured, they had more class, more civility, than any American I knew.

Implicit in Hill's treatise is the morally persuasive invocation of fear as an impairment of vision and, therefore, as a barrier to social change. By way of the Vietnam example, he crafts an image of a being whose socially acquired resistance to the Vietnamese people is peeled away as he is released from fear's grip. Only upon that release is the elimination of bigotry rendered possible. The cognitive structure embedded in this fear scenario is relatively straightforward:

Vietnam
(event)
↓
Fear
(emotion)
↓
Social action/critique
(promotion of change in oneself and therefore change in society)

Fear and Moral Meaning: Loggers

When loggers speak of fear, they are most often talking about the acute dangers inherent in their places of work. Fear and discussion of job-related dangers dominate their conversation as their form of employment continues to have one of the highest occupational mortality and injury rates. Experienced loggers are quick to recognize viable crew members, those willing to exert themselves to secure their co-workers' safety. Nonetheless, accidents are frequent, leaving an uneasy pall over most worksites – a pall that must be emotionally managed. Talk of fear emphasizes the need to suppress one's consternation so as to maintain control over ever-present dangers. According to Jim Stratton:

Well the logger couldn't admit that he was fearful except to one another ... but there's a lot of fear, yeah, because there's always in the back of your mind that one moment of danger. That could mean your death is there just like that [snaps his fingers], you know. There's often nothing you can do to prevent it no matter how safe a worker you are, how conscientious you are, how safe you might [pause] it's still there.

Mark Evans talked about his "worst fear" coming true the day his son Glen was injured on the job. A felled log, he recalls, struck a standing "wildlife" tree, changing the log's anticipated trajectory. The runaway timber came at Glen (who was not permanently injured) "like a freight train and sent him flying about 60 or 70 feet."

Both Jim Stratton's and Mark Evans's fear scenarios reflect the very real hazards of logging. But they also hint at another sentiment and tell a particular moral story. Specifically, loggers resent the fact that they work under incredibly dangerous circumstances in order to supply an often-critical public with a desired commodity. In their view, they trade danger and commitment for the insult of finding themselves stigmatized as environmentally destructive. Notice, for instance, that Evans makes a point of mentioning the "wildlife" tree that was involved in the generation of the runaway log in the first place. Federal policy pertaining to cutting practices on public land requires that a few wildlife and seed trees be left standing in all clear-cuts. They are said to provide opportunities for animal shelter and forest regeneration. The futility of expecting one standing tree to house many species and reseed multiple acres aside, "wildlife" trees symbolize regulations that are widely regarded by loggers as the by-product of the meddling of environmentalists. In the logging-controversy context, talk about fear (and danger) thus becomes talk about unrecognized sacrifices made by loggers. The embedded emotional scenario is also straightforward:

Logging accident
(event)
↓
Fear
(emotion)
↓
Association of physical injury with social sacrifice
and therefore social critique
(social action)

Chris Baker talks of fear, danger, and the death of his friend. He confirms the scenario hypothesis, offering, in fact, a more extreme example in that he actively and consciously blames environmentalists for occupational injuries.

I had a friend killed last year by a wildlife snag that they left. They had to leave. And it – they was working on this unit, and this – and the skyline – the skyline caught the edge of the snag and it come down on him. It was an unstable snag. You know, it was ready to fall at any time, but they had to leave it, because that was – that, you know, because of regulations, that was a wildlife snag. And it killed my buddy. You know, I'm very bitter about that, you know, because I think of guys like Matthew Waite and stuff. My friend's dead because of people like him.

In the end fear and danger are clearly about physical survival in the woods. But fear, too, communicates a moral claim; it captures loggers' sense of their assigned position in the social universe. Loggers see themselves as contributing to a productive society that demands wood; consequently, they use emotional language to criticize those who stigmatize them and underestimate their utilitarian contributions.

Emotions and Ethical Practice

It should by now be clear that activists' references to fear are principally focused upon possibilities for social critique (loggers) and social change (environmentalists) and that, in this sense, they uphold the premise that emotional language is a morally persuasive medium. It is also the case that the moral tenor behind such language helps to propel the speaker towards not just broadstroke positions (such as the need for social change), but also towards visions of specific ethical practice. Ethics, in this case environmental ethics, are defined as putting into practice notions of right versus notions of wrong conduct towards nature (Armstrong and Botzler 1993; Proctor 1996; Rolston 1999). Further, their link to morally endowed emotions is basic to the extent that how one conducts oneself in nature (i.e., how one values nature) is "grounded in human feelings" that are "projected onto the natural" world (Callicott 1984, 305). Emotions, for philosophers and anthropologists alike, are inseparable from, and indeed disclose, moral values (Stocker and Hegeman 1996).[11]

Among the ethical divisions central to the old-growth dispute is that between anthropocentric and biocentric positions (Callicott 1995; Norton 1991; Rolston 1994, 1999). An anthropocentric ethic posits that nature's worth is derived primarily from its capacity to serve human ends. A biocentric ethic respects all living organisms: because nature is alive, it is regarded as "good" in its own right and thus deserving of moral consideration. More radically, biocentric deep ecologists (Devall 1992; Naess 1989) seek to erase the line between human subjects and natural objects. Modelled after ecology, they promote the interconnection and interdependence of all life.

The anthropocentric/biocentric distinction corresponds roughly to the respective positions of loggers and environmentalists.[12] For the latter group, ethnographic illustration of the link between fear, the moral necessity of cultivating a self or being open to change, and ethical practice is readily manifest in Donald Richardson's exposition on the necessity of an emotional attachment to the Earth. Two middle segments were omitted from his above-cited passage; they are now recorded below. The first segment speaks to the deep ecology premise that the individual self should be ecocentric rather than egocentric and, thus, indistinct from the rest of the living natural world. The second segment links this ethical position to emotion, specifically fear. Richardson's ideal self resonates with his earlier mentions of emotionality, wherein the self ought to be alive, exposed to the cruelties and wonders of life, so as to break down humankind's fear-driven resistance to the interspecies bond that connects life on Earth.

Taking up his above-quoted passage from "I'm an animal and proud of it," Richardson continues:

> I'm an animal and I'm proud of it. The oceans and the earth run through my veins. The wind fills my lungs. The mountains make my bones. When a chainsaw rips into a 500-year-old Douglas fir, it's ripping into my guts. And when a Norwegian-powered whaler fires an exploding harpoon into a great whale, it's my heart that's being blown to smithereens. And when a Japanese bulldozer rolls through the Amazon, it's my body that's being ground up with all the soil organisms. Because I am the land and the land is me and I care about it. And we have got to somehow form an emotional attachment with this Earth, and sink our roots deep into it.

Thereafter, in the segment that precedes his closing "we've got to be emotional, we've got to be passionate," he adds: "In this modern age we're so *afraid* of death, we're so *afraid* of pain that we build walls in front of ourselves. We're *afraid* to open ourselves up to love another person honestly because they might hurt us. We're *afraid* to fall in love with a piece of country because the next time we go back it might be a clear-cut, and so we wall ourselves off."

The moral claims of environmentalists regarding the need for a society of beings open to change comes together here with Richardson's representation of his human self as indistinguishable from the non-human world. When he claims that the chainsaw tearing at the fabric of an old-growth Douglas fir is tearing at his own body, one can envision Richardson peeling back his outer layers of skin and quite literally embodying the threatened tree. Concepts of emotion and self are closely intertwined in that emotion terms invoke the speaker's concept of how oneself and others "ought" to feel and thus act. Moreover, Richardson's concept of the self is unique in that he identifies principally with the biotic community and not the human community. In this sense he is calling forth a new self that directly contradicts social scientists' findings that self-claims are, generally speaking, either sociocentric or egocentric (e.g., Kusserow 1999; Markus and Kitayama 1991).

Studies of self and personhood have indicated that all human cultures distinguish what we might call the self – the idea that we possess subjective perspectives, have self-images, which reflect our position-specific (and culturally influenced) theories about why we behave the way we do and how we should behave towards human others (Kusserow 1999).[13] A sociocentric self has often been associated with non-Western, particularly South Asian, groups who are said to experience a more socially cognizant self, a self defined by social role, family, lineage, and so on. In some Asian cultures emphasis on the collective good is said to heighten an individual's sensitivity to the needs of others and lead to behaviour that crafts an interdependent and thus very other-conscious self. In contrast, the egocentric self is most associated with "the West," wherein the importance of individualism leads to a self that is comparatively unaffected by the behaviour of those around it and, thus, is construed as a bounded, separate, or autonomous unit (Joseph 1993, 9). This West/non-West distinction is likely a gross oversimplification.[14] But the point here is that all research on the self points to a human-centric bias wherein talk of the self speaks primarily to the degree to which an individual identifies with members of the human community based on either autonomous/egocentric or interdependent/sociocentric definitions of the self. It is a bias that Richardson's behaviour patently rejects.

The work of deep ecologists (Devall 1992; Naess 1989) and, more recently, ecological phenomenologists (Abram 1996) has provided a powerful argument upon which more radical environmentalists draw in the hopes of fundamentally altering the degree to which humans identify with non-human communities. Among their goals is the effort to redefine community to include non-human others. When Richardson says, "I am the land and the land is me," and when many others speak of cultivating an emotional attachment to the Earth, they reflect deep ecology's extension of the ecologist's credo: the interconnectedness of all things. Deep ecology posits that humans as well as all other species manifest their "true" selves only as they realize their connectedness to the natural environment. An "authentic" ecological self is generally sought through wilderness experiences, poetry, song, dance, and ritual practice (Taylor 1993). The emotional impact of song and dance, Durkheim's (1965) "collective effervescence," transports participants to a different emotional plane and, it is hoped, pushes them to question at every level the dualist assumption that humans are separate from and/or exist in opposition to the natural world. The implication for the human self is that, rather than thinking or operating as a bounded, autonomous unit – or even as an interdependent socially oriented unit – all firm boundaries between the self and nature must be dissolved. Writing on the teachings of deep ecology, Rolston (1999, 415) puts it thus:

> The human "self" is not something just found from the skin-in, an atomistic individual set over against other individuals and the rest of nature. Rather the "self" is what it is with its connections; the self takes up its identity in these interrelationships with the biotic community ... Ecology does not know as encapsulated ego over or against his or her environment. Ecological thinking is a kind of vision across boundaries ... The human vascular system includes arteries, veins, rivers, oceans, air currents. Cleaning a dump is not that different from filling a tooth ... metabolically and so metaphorically.

Deep ecologists have themselves noted that humans do not easily embrace the principles of deep ecology as they defy so fully the more familiar and thus socially comfortable separation of the human and animal worlds. Others argue that activists need not assign personhood to all non-human species but that they can nonetheless operate more fully in an intersubjective perceptual field (Milton 2000). The central goal for this later, ecocentric self is to remain cognizant of, and more fully attuned to, the "more than human world" (Abram 1996).[15]

Loggers and the Anthropocentric Self

Given the radical and fundamental reorienting of human life indicated by a realized ecological self, one should expect that the emotional discourse of loggers deeply and passionately opposes this newly configured self. Indication of their opposition to the general premise that concern for humans should supersede concern for nature is evident on numerous occasions in earlier chapters and is often simply stated as their hope that "society will resist a growing tendency toward emotional interactions with nature."[16] Loggers' comments support the many scholars who have cast them as adhering to an anthropocentric ethic, as promoting, within the context of land management, a sociocentric self concerned with the well-being of forest-dependent communities (Porter 1996; Proctor 1996). Loggers, they argue, are willing to protect nature to the extent that such protection conserves nature ad infinitum for human ends. Nature's value is derived from its sustained capacity to provide for material needs, generate employment, and/ or support rural communities.

Evidence of the depth of loggers' resistance to deep ecology, and the emotional language used to convey that resistance, often surfaced during public speech events. Dan Marden is a dominant presence at pro-timber conference events. For several years running he spoke at the annual gathering of gyppo loggers, small contractors, and small independent mill owners. Both Richardson and Marden provide examples of the change-inducing operation of emotional language and behaviour. But while Richardson attempts to compel his audience into redirecting their self/other orientations, Marden relies on the need to fear rather than embrace change, along with the combined effect of moral shocks – jolts aimed at emphasizing the moral affronts faced by his audience (Jasper and Paulson 1995).

Marden, like Richardson, draws concepts of fear into segments of his talk. But his use of the term is opposite to that of Richardson. He tells his audience that as a young member of the US Marine Corps he witnessed the raising of the Berlin Wall. "I saw," he declared, "the wall, the wire, and the weapons. But more importantly, I saw the looks of fear and terror on the faces of Eastern Europeans and I knew in my heart that the wall would never come down. And yet it did." Marden is the only timber advocate to use fear to refer to something other than the moral claim that loggers suffer morbidity and mortality in order to deliver to the public desired material goods. He upholds fully the opposing (and dialogic) structure characteristic of timber and environmental activists' emotional language; he mirrors environmentalists' fear-resisting call by invoking its opposite. In Marden's talk "fear and terror" refer not to the opening up of the self to embrace change but, rather, to the necessary shutting down of a people as they face an oppressive state. Change, in this sense, is said to be terror-inducing; in fear's face, the only logical response is retreat. Despite this retreat, Marden adds, hope is possible: When faced with abhorrent change, "the wall" did, after all, "come down."

Marden follows his reference to fear by dismissing environmentalists as "disaster rhetoric" zealots who erroneously claim that "the sky is falling" and that "mankind is a cancer on the planet." Thereafter he offers a series of morally shocking vignettes, each of which is organized around the theme that small landowners and ordinary working people are being crushed by federal regulatory policy. In all of his cases, the protection of habitat is said to be placed ahead of the individual's right to earn a living from the land in question. Playing off classically American themes of individual freedom, economic liberty, and freedom from government intervention, the tempo and volume of his exhortation escalate. Finally, he posits that "Western civilization" is itself threatened.

A prominent emotion in and around social movements, moral shock – the ubiquitous "can-you-believe-what-they-did" claim – helps stir up a sense of shared outrage and thus creates an affective opening into which the target of one's indignation can be situated (Jasper and Paulson 1995). Marden uses that opening to admonish his listeners for tolerating a deeply ecological ethic while simultaneously assigning to them the appropriate emotional response – the need to be "mad as hell."

> Ladies and gentlemen, at a time when snails are more important than people; ladies and gentlemen, at a time when birds and fish and plants and trees and flies and spiders are more significant than men and women, their ability to provide for jobs and community and family and heritage, to live with dignity and respect, I hope you're with millions of Americans, I hope you're *mad as hell* and you're not going to take this any more!

Admittedly, I find Marden's conservative rhetoric disturbing, as do many others (Brick and McGreggor Cawley 1996). His talk, interestingly, received strong though not overwhelming applause. (Afterward, a few loggers I knew approached me, eager to tell me that they thought Marden was too strident and had overstated his case for dismantling regulatory authority. Careful management of limited resources was, they agreed, necessary. These comments appeared entirely consistent with their efforts to contain emotion and remain "reasonable" in the face of controversy.)[17]

Audience response aside, it is important to note the interplay of fear, anthropocentrism, and moral shock in Marden's speech. Having assigned his audience the right to fear (and retreat from) inappropriate moral change (which he defines through the equation biocentrism = totalitarianism), Marden moves into the opening that moral shock has provided and asserts his vision of the supremacy of the human position vis-à-vis the non-human world. The utilitarian needs of humans are pitted against the importance not of keystone species – spotted owls, grizzly bears, or salmon (or to what environmental ethicists like to refer to as charismatic megafauna) – but, rather, to generic small animals, "pests" (flies), and the kind of invertebrate that often inspires phobias (spiders).

ᔑ

In January of 2001, a large timber company in the province of British Columbia announced that its logger-employees would have to undergo sensitivity training in order to avoid confrontations with environmentalists (Hamilton 2001). It was a move entirely consistent with the opening pages of this chapter, which refer to the oft-noted observation that the forest dispute has become too emotional. Excessive emotionality had come to mean, within the Oregon context, that loggers must contain their rage in order to counter their public images as irrational players and that women were "closer to nature" than men. Both positions can be viewed through the lens of emotion scholars who posit that being emotional is not simply a behavioural realization of how activists feel but, rather, a clear expression of the structural positions assigned, respectively, to working-class loggers and women in American society. Yet a greater emphasis on identity and agency soon revealed that such positional assignments can be resisted. Loggers can and did protest the need to dissociate themselves from expressions of emotion; but they spent equal or greater time emphasizing the price they had paid for their muted affective front. Their emotional caution, they claimed, had betrayed them by concealing the true depth of their despair. Environmentalists, on the other hand, demonstrated no particular need to control their emotions and, instead, used invocations of emotion to promote an intersubjective space within which a primary bond could be established between humans and wild nature.

Thinking of emotional language as a morally communicative medium added significantly to this portrait. Acknowledging that the affective language of activists marks identity movements as moral movements, as collective action that helps individuals articulate their moral vision, made possible the tracing of different ethical practices that emotional language implies. It thus became evident that affective language was a central medium through which activists promoted a particular self vis-à-vis the non-human world. Presenting fear in moral terms indicated that loggers equated fear with the experience of physical danger at work and the misrecognition of their social contribution. Further, loggers equated fear with resisting change to the improved status of the biotic community in the moral universe. Environmentalists, conversely, equated fear with society's failure to embrace social change and to cultivate an ecocentric self more fully open to, and indistinguishable from, the non-human world.

8

A Concluding Discussion: The Triangular Shape of Cultural Production

This book began with a single, guiding question: In what sense can the dispute over the Pacific Northwest's temperate rain forest be said to be a cultural one when the debate is phrased in terms of, and so appears to be driven solely by, disputes about legal, scientific, and land-managerial priorities? Early on in my endeavours those I met would respond to the prospect of a cultural study with a certain delight. "Oh, it's about time," they would say, or "thank goodness, we really *need* you!" These responses appeared, in some cases, to be a plea from loggers to have me fulfill their wish to become a recognized people. On other occasions, for both loggers and environmentalists, the point seemed to be that the enormous attention granted the science or economics of spotted owl survival failed to capture the depth of the perspectival differences at play. The unnerving impression was that I could somehow explain the most "innate" properties of the collective psyche of loggers and environmentalists and thus shed light on a battle that had paralyzed a region characterized by a local politician as the most divided constituency in the country.

Well, I have not described any innate properties, at least not to my knowledge; rather, it should be apparent that one cannot, to put it colloquially, "pin down" the beliefs of loggers and environmentalists like dead butterflies on the naturalist's display board. One can only conclude that the differences between groups are profound, morally rooted, and ethically challenging. Moreover, there is only, at best, some patterned logic to the different ways in which loggers and environmentalists imagine themselves as living in and behaving towards nature. Briefly and generally stated, loggers see themselves as members of historically rooted land-based communities whose experiential knowledge of the forest is sound and wise but who have nonetheless been cast unfairly as violent antagonists and treated without respect, despite their wood-producing contributions to society. Conversely, many ancient-forest activists lean towards, and derive insight from, cultural arrangements that they imagine as resembling Aboriginal practices.

They are wary of, though they also endorse, a science that stands metaphorically for nature as mystical, complex, enchanting, and vulnerable to disruption (Yearley 1993); and some are creatively resistant to emotional norms that interfere with a deeper bond between the human and biotic worlds.

Despite these patterns, individual loggers and environmentalists express their agency – and their creative play with dominant discourses – differently. Amidst the struggle to maintain a voice in decisions about an already depleted land base, some activists from both groups are visionary, some are vindictive, some are prejudicial, some are none of these. Moreover, Oregon's loggers and environmentalists are ingenious beings. They invent, elaborate, and portray effectively their respective group identities in relation to key topics and practices in American society. They draw from and play off the popular ideologies of democracy, science, cultural authenticity, and the normative rules for emotional comportment. Their varying discourses and behaviours are acutely sensitive to the power of these enduring cultural discourses and yet also fully attuned to harnessing them in order to promote their own versions of the cultural world of human-environment relations.

Understanding Environmental Conflict as Cultural Conflict

The most significant lesson I have learned in this study of environmental conflict is this: it is crucial to explore both the tension between the variations in activists' strategies (as realized through the medium of identity) and the commonalities in their conceptions. This tension between variation and commonality is a central site of culture in action. I could only have seen this by weighing equally, and considering in the same ethnography, the challenges put forth by loggers and environmentalists engulfed in the cultural world of western Oregon in the 1990s. Thus, my conclusion turns to this dual-group, dominant culture-challenging method of study and what can be learned by examining activists' competitive dialogues about cultural forms – specifically, those pertaining to emotion, authentic community, political legitimacy, and the science of forest management. Thereafter, I will explore how the implications of this study may help policy analysts understand how the public values nature and, thus, better represent public thinking about land management decisions.

The Triangular Shape of Cultural Production

There are many superb ethnographies of agrarian and resource communities, of those who struggle within powerful regimes to farm, log, fish, drill for oil, or tap for rubber (e.g., Brown 1995; Carney 1996; Carroll 1995; Grossman 1998; Hecht and Cockburn 1989; Palsson and Helgason 1998; Sheridan 1988). Several similarly astute studies of environmental activism (e.g., Berglund 1998; Milton 1996; Strang 1997), new social movements (Jasper 1997; Laraña, Johnston, and Gusfield 1994), and environmental justice movements (Johnston, 1994, 1997; Szasz 1994) are available.[1] But as ethnographers of environmental and land-labouring communities, we have generally failed to grapple with the position of more than one party simultaneously, except when the two parties are clearly linked to one another in super- and subordinate positions. The overwhelming majority of environmental ethnographies remain preoccupied with (1) single groups, (2) single groups who contest superordinate power, or (3) studies of single groups involving second or multiple parties on a peripheral basis only.[2] The propensity for anthropologists studying environmental conflict to take up the plight of one group or one cause has hampered our ability to focus on the many creative, engaged dialogues *between* subordinate groups – dialogues about nature, cultural meaning, resource use, and future practices. Our abiding commitment to solo-group or solo-movement studies ignores the lively cross-party identity-shaping defining of difference that is taking place and, thus, the triangular structure of environmental conflict in much of North America and beyond. In other words, we ignore the mutual (and dialogic) orientation of one group to the other and of both to the more powerful discourses and enduring cultural forms that infuse their exchange.

Triangular studies of conflict such as this one recognize that there are often two or more groups (more or less marginally positioned) vying for the right to define land management decisions. Viewed together they reveal the diacritical disturbance (i.e., the two "bottom" points of the triangle) aimed at the larger cultural systems (the "top" point of the triangle) of which land management regimes are a part.[3] Such studies follow Marcus's (1998) claim that an emphasis on encountering conflicting parties is increasingly appropriate to the study of contemporary society. Conflict is becoming, he argues, a central organizing principle for multi-sited ethnographies (i.e., those focusing upon more than one "people" or place). The very few studies of environmental conflict in which multiple marginalized groups compete to contest and produce cultural forms have been extremely productive. Rodríguez's (1987) early study of tension over land and water in Northern New Mexico is a case in point. She encountered a tri-ethnic trap wherein Hispano farmers and Pueblo Indians (alongside a White activist support base) were forced into an oppositional bind. Each group was forced to differently articulate its right to cultural survival due to its respective legal position vis-à-vis state recognition of land rights. More recently, Gezon (1999) documents inter-ethnic struggle as endemic to a dispute over marine resources in northern Madagascar. In her study, ritual innovation and enactment are employed by an ethnic minority in northern Madagascar to offset the control of the resources by the region's ethnic majority and by encroaching international groups.[4]

A triangular approach to the study of environmental conflict also fosters anthropology's effort to move away from theories that reify a dichotomous relationship between nature and culture. Historically, nature was seen as a "thing," an a priori environment to which humans adapt (Biersack 1999). Interest in poststructuralist theories subsequently highlighted the fact that there is no such thing as a prediscursive nature (Descola 1996; Escobar 1999; Strathern 1980). That a material "natural" world exists is a given. Nature is, nonetheless, constructed to the extent that it is always being labelled and thus defined as natural capital (for economists), as biologically significant (for wildlife biologists and ecologists), as morally good (for ethicists), and so on. More recently, attention is being paid to both the human-to-nature interactions that produce different environmental spaces (wildernesses, refuges, parks, swidden through to industrial agricultures, etc.) and the nature-to-human interactions that affect society. The attention to both directions is what political ecologists refer to as striking the balance between the "social construction of nature" and "the natural construction of the social" (Watts and Peet 1996, 263).

In Oregon, activists work to reshape (or construct) nature in their own imagined vision, but their thinking about nature, or, more accurately, *with* nature in mind is also the impetus for new social patterns (i.e., is an example of how nature constructs the social). Close scrutiny of the negotiation and conflict processes between protagonists concerning what nature *is* throws into relief both the emerging possibilities for a new nature and the impact of that dialogue upon the social groups involved. Two examples from the old-growth dispute illustrate this dialectical nature-and-society-redefining point. Loggers, as set out in Chapter 4, became a self-conscious community, a new grassroots social group defined by their understanding of, and claim to, forestland. They did this in response to the rise of locally unfamiliar cultural definitions of the natural as rare, ancient, sacred, wild, and biologically indispensable – definitions introduced by the wilderness, environmentalist, and scientific communities working to save old-growth forests. In Chapter 6, the negotiation over what constitutes a legitimate place-based community was paramount. Just whose notion of ideal land-based communities will prevail is not yet clear. Nonetheless, it is apparent that many environmental activists are vying for an Aboriginal-centric model, all the while working to diminish (via claims of loggers' rootlessness) society's recognition of loggers living in unique, land-based communities. Loggers, meanwhile, are struggling to support a vision of the forests as working and workable land by invoking their status as one member in a long trajectory of communities (Aboriginal and non-Aboriginal) with intimate knowledge of the land they labour.

Cultural and discursive analyses highlight these constructive processes; but such analyses are not yet fully triangular or dialogic at the subordinate level. This has implications for theory in environmental studies and political ecology. Discourse analysis, with the tradition of recent critical theory, draws our attention to how marginalized groups contest entrenched cultural discourses by emphasizing some concepts at the expense of others and by signifying some meanings as more important than others within the context of change-directing or change-contesting social action. The focus is generally on the tension between hegemonic discourses and competing subaltern discourses. Peet and Watts (1996) clarify the relationship between dominant and subordinate discourses, following Maffisoli, by offering the useful distinction between hegemonic discourses and environmental imaginaries (a distinction that also helps avoid the tangling language of capital "D" hegemonic discourses versus lower-case "d" subordinate ones). Dominant discourses are generally affiliated with late capitalist institutions (especially modernist economic and agrarian development schemes), which are recognized as relatively stable formations. Imaginaries are alternative ways of imagining nature; they are expressed regionally and include "those forms of social and individual practice that are ethically proper and morally right with regard to nature." Imaginaries coexist uneasily against the stable (hegemonic) forms; they are the discourses of "competing, even conflicting, cultural, racial, gender, class, regional, and other differing interests" (Peet and Watts 1996, 14). They have emerged as a "primary site of contestation; critical social movements have at their core environmental imaginaries at odds with hegemonic conceptions" (263).

In most studies, however, the relationship between imaginaries and dominant discourses remains a two-tiered hierarchical one. Because multiparty ethnographies of competing imaginaries within dominant cultural frames are rare, the implications for understanding and writing about environmental movements are only partially realized. Ideally, we need to place greater emphasis on multiple competing activists and their creative play with master discursive categories. How does cultural production take place and under what circumstances does it alter or generate a new dominant discourse? How are master concepts worked into different contexts? In other words, how are concepts like "community," "cultural authenticity," or "rights" amplified to build an image of stable long-standing group identities of people who deserve a stake in deciding land management practices (Brosius, Tsing, and Zerner 1998)? And how, and under what conditions, do protest communities succeed (or not) within larger land management schemes?[5]

We recognize the agency of activists but we have only the vaguest sense of the long-term implications of that agency or how best to capture cultural production in the making. I can see that loggers are promoting a common sense approach to science (Chapter 5) or are playing off of the experience of stigmatization (Chapter 4) to ensure that their voices are heard, that they are recognized politically and socially. Simultaneously, I can see that environmentalists are working to dismiss loggers as legitimate spokespersons for the environment by casting them as puppets of industry and that they want to promote the mystification of, and enchantment with, nature. But I do not understand how the dialogic exchange of activists concerning science or grassroots legitimacy plays out in the larger scheme of things. It is more than random coincidence that stakeholder groups – representative of industry and government, NGOs, and community groups – that gather to discuss and resolve environmental disputes are more popular than ever (Brosius, Tsing, and Zerner 1998; Gregory 2000). Resource workers are increasingly welcome members at decision-making tables. (And it is telling that some activist loggers, along with key environmentalists, were included in President Clinton's Forest Summit.) This suggests that the growing agency of resource workers has been influential at the level of policy. But environmentalists' popularization of the more abstract and mystical definitions of nature as linked to science, or the idea that our thinking about community should be extended to the non-human world, has also been influential. Surveys of both timber-dependent and non-timber-dependent communities reveal increasingly widespread (and mutually agreed upon) support for the once radically green idea that "all species, including humans, have an equal right to co-exist on the planet and that humans are attracted to the spiritual qualities inherent in the natural world."[6]

Borderland/Zones of Friction
A triangular model of ethnography might also fulfill Ortner's (1999, 8-9) call to reconfigure studies of culture by (1) paying close attention to the intersection of cultures, the borderlands[7] or zones of friction wherein groups meet, and (2) situating cultural analysis within and beneath larger analyses of social and political events and processes. Her suggestions reflect the doubt contemporary anthropologists have cast on the idea of culture as a "consistent" way of being for all members. "Such descriptions have too often made it sound as if all of [culture] X thought, felt, and acted the same way" (Strauss and Quinn 1997, 3). The speciousness of these descriptions was brought fully into focus as ethnographers, following Foucault (1978, 1979), investigated power and its implications for how those with, and particularly without, power experienced, understood, and operated within their "host" cultural systems. The questions of what a cultural system (its religion, its ideology, its science, its nationalism, etc.) *is* (and how it affects whom) emerged as too important to support "thick descriptions" of other people as an end in itself. Rather, argues Ortner, ethnography should be a means for illuminating struggle and change (i.e., agency) within systems of power. It is necessary to draw out cultural difference within these zones of conflict so that "the clash of power, meaning, and identities" is revealed (8).

Each of the preceding chapters has situated cultural analysis within and beneath larger political events and, thus, has highlighted how loggers and environmentalists compete between one another while being subordinate to dominant systems of landscape definition (e.g., scientific ones), political process (e.g., grassroots legitimacy), appropriate emotional bearing towards social and biotic communities, and dominant definitions of what is and is not an authentic land-based community. The emphasis rests squarely on the creative and innovative behaviours taking place within zones of friction as grassroots competitors act in concert with one another. In so doing, what comes to light are the particulars of how different identity groups vie with one another while being influenced by the same entrenched cultural forms.

Recall, for instance, Chapter 7's discussion of how protagonists differently managed the normative expectation for, and therefore constraints upon, emotional control. In that chapter, loggers and environmentalists played with the leeway allotted by the many possible definitions of emotion as a master cultural construct. It was noted that, while emotional control in the form of cool reason and civility is often aspired to in North American society, one can also appeal to the necessity of emotion as passion and, thus, the basis for a uniquely charismatic freedom from the constraints of society. Both activist parties worked within the shackling constraints of this bi-polar valuing of emotion. In order to come across as having risen above class-affiliated, socially "inappropriate" bursts of anger, loggers conscientiously and poignantly promoted themselves as emotionally contained. At the same time, they actively questioned their failure to derive significant benefit from their emotional obedience. They used that experienced failure to critique society for misrecognizing the depth of their collective despair. They also used the morally persuasive language of fear to promote utilitarian uses of the forest over and above ecocentric ones.

Environmentalists, due to their middle-class and gendered positions, embraced emotionality in order to conjure up the possibility of a world in which one "relates to nature." They used emotional language, especially the language of fear, to promote a world in which human beings open themselves more fully to the non-human world. Embedded within this dance between environmentalists' and loggers' portraits of emotion is the possibility that either might become the prototype for appropriate human/nature interactions, the basis for new cultural forms. Ultimately, this is an interactive and dominant system responsive (and sometimes producing) dialogue about which we still know very little.

Balancing Cultural Resistance and Cultural Production
My intent is not to posit a simplistic model of cultural change along the lines of: activists make noise, introduce new language, contest one another's

portraits of science or cultural authenticity, and therefore change the ideational world within which they are situated. The interplay of macro- and micro-level forces that produce change are undoubtedly more complex. And the dialogic processes of identity formation and evolution are, by definition, "moving targets" (Johnston, Laraña, and Gusfield 1994, 16). They do not avail themselves of a linear, predictive theory in which A (activists) precedes B (action) and thus produces C (cultural change). One also runs the risk of exaggerating "the effect of protest movements" by assuming that their "innovations will be universally adopted" (Jasper 1997, 374).

My concerns about theory and ethnography in environmental studies lie elsewhere: they lie with the preternatural schism, in studies of protest movements, between resistant and productive processes. Resistance studies are concerned principally with how individuals (and the identity-centred environmental movements with which they are associated) defy political and ideological oppression. Within environmental contexts this may include studies of peasant revolts aimed at reclaiming usurped agricultural lands, local opposition to state-mandated closures of fisheries, or farmers' resistance to globalization policies that open local markets to internationally produced foods. In extreme conditions such resistance emerges as an oppositional style, as people work to eke out self-determination in near-to-impossible circumstances. These acts of defiance are often creative, but the weight of circumstance is such that the defying agents remain trapped in and even ensure (i.e., re-produce) their continued oppression.[8] Studies of cultural production emphasize, alternatively, the ability of groups to creatively proliferate new ideas, new ways of being towards nature. The emphasis is overwhelmingly on the generation of alternative and often novel practices that may one day become the basis for new cultural institutions.

The production (protagonist)/resistance (antagonist) distinction is unproblematic until we begin seeing both forces at work in one study, particularly when that study involves environmental activists and natural resource workers. When groups are placed together within the same matrix, the ability to label one group as "producer" and another as "resistor" affords the ethnographer unwarranted power. The group labelled by the ethnographer as the protagonist party becomes (implicitly) the hero(ine) in the overall conflict drama – the most important entity, the baseline, against which all other action is defined. The antagonist-resistor is, by comparison, pitiable – an opposing force against which the more valued and central producer-protagonist position is evaluated. Or the resistor's position, in the case of natural resource workers, is rendered indistinct from the more powerful industry forces with which it is associated. Lange (1996), for instance, draws on communications theory and frame analysis when writing of the spotted owl dispute. I agree fully with his finding that both parties vilify their opponents while ennobling themselves, that both parties manipulate "facts, explanations and interpretations ... to discursively construct a reality favourable to one's rhetorical goals" (139). But his parties are distinguished as protagonists (grassroots environmentalists) and antagonists (industry representatives); loggers are included only as epiphenomena (the now familiar "pawns") of industry. Thus the agency of loggers is suppressed, as is the latent class position of many resource workers. (I say "latent" because the language of logger identity has come to replace the language of class.) But class does not go away. As theorists, we should problematize this invisibility, not ignore it, however much we might disagree with some of the more conservative viewpoints put forth by some (though certainly not all) resource workers.

We should also be careful not to conceal the advantages of privilege and the disadvantages of class-based injustices. Environmentalists protesting resource extraction are marginalized as activists yet privileged in that they need not demand for themselves the right to job security or political representation. They pursue, instead, the protection of others and of other species (Jasper 1997). We should ask ourselves whether that privilege grants environmentalists room to manoeuvre as culture producers. Does imagining the possible coincide more often with those less affected by structural change? The only possible answer in this context is: Yes, for the most part, but not always. Loggers, and other resource workers, shunned overt class arguments for identity and rights-based claims and thus managed to build a formidable social movement that capitalized on public sympathy for the loss of the logging way of life. Consequently, they gained purchase in the conflict despite the fact that a traditional ally, the labour movement, was relatively moribund at the time. But, more frequently, loggers came across as defensive and anachronistic, whereas environmentalists come across as visionary, as capable of expansive thought. In Chapter 7, for instance, loggers resisted the necessity of emotional control, but they had comparatively little to say about new "emotional" ways of being or expanded (biotic + human) definitions of community. Their position was, as resistors, a defensive one. Similarly, in Chapter 5, loggers adhered – in an identity-defensive manner – to the "dated" idea that forests are a crop, while environmentalists played with the mystical implications of the complexity of forest structures.

One important means for achieving a synthesis of productive and resistant processes and, thus, for comprehending where and how each flourishes, is to rely less heavily on assumptions about political domination since they can, to some extent, corner theorists into choosing as a focus a single laudable group that best embodies activist (and often academic) ideals about local forms of resistance. A wider ethnographic focus, including an in-depth look at the conflicting social movements engaged in struggles over land and resources, is warranted and can be accomplished through analyses that include "all of the parties to the struggle, with their different perspectives on and stakes in their interdependent lives" (Holland and Lave 2001, 23). The interlaced vertical and horizontal connections between and across social actors is undeniable. As such, it is insufficient to frame the social and political analysis of natural resource disputes as, primarily, a struggle between hegemony and resistance. This would only serve to preordain the boundary and characteristics of both the struggle and the participants involved and, therefore, would overlook the rich and disputatious cultural contest in which "multiple, diverse, and interconnected" parties are engaged (23).

Policy Implications

Those who are concerned that policy includes a wide array of social perspectives must also be concerned with how (and under what conditions) environmentalists and loggers can be positioned so that their creative and productive perspectives can be understood. Conflict over scarce natural resources will not be "solved," and improved logging practices will not be created if the more imaginative and experientially wise activists on both sides are silenced. This last point leads directly to questions about the development of forest policy. Increasingly, regulatory agencies (the Forest Service, the Bureau of Land Management, the Environmental Protection Agency, the Fish and Wildlife Service, etc.) responsible for land management have turned to public opinion to determine socially and ecologically defensible natural resource plans. These efforts are commonly known as value-elicitation processes; they use small group ("stakeholder") discussions and surveys to ascertain how the public values the environment. This trend has been lauded for its democratic underpinnings and its move towards synthesizing "lay" and expert perspectives. It is a trend I generally support. But several features of this study – namely, the predominance of activists' moral concerns and the situational uniqueness and context specificity of many of their imaginings – point to the difficulty of developing sound value-elicitation practices. Two key considerations are in order: the first concerns language, power, and creative thought; the second concerns the problem of pinning down stakeholder values.

Language, Power, and Creative Thought

The practice of eliciting stakeholder values concerning nature is currently dominated by economic approaches, despite the infusion of social, psychological, and anthropological approaches as well as input from the new subdiscipline of philosophy known as environmental ethics. Predominant among valuation techniques are cost-benefit analyses, including contingent valuation surveys of what the public would be "willing to pay" in order to improve the status of a forest, a spotted owl, or a national park. Embodied in these approaches is (1) the assumption that the majority of the public endorses rational, economic expressions of the hypothetical market-value of nature and (2) the assumption that monetary expressions of value reflect that which is held dear, worthy of protection, and ethically or socially esteemed.

Yet, it should be apparent that such dollar-centric studies "inscribe" onto stakeholders and the land particular forms of discourse. They create "certain ways of thinking about nature and preclude others, privilege certain actors and marginalise others" (Brosius 1999b, 36). The trouble is that the language (in this case an economic one) used to discuss stakeholder "values" determines the values that are "allowed" to surface.[9] The problem is an important one because the imposition of an economistic model contradicts the fact that much of what stakeholder loggers and environmentalists have to say concerns moral and ethical meaning and not economics per se. The "paycheque" certainly matters, but it does not begin to convey what, for loggers, is at stake. Similarly, Brosius found that, in a Malaysian context, the institutionalization of stakeholder processes insinuated and naturalized a discourse that excluded moral or political imperatives in favour of bureaucratic and technoscientific ones. This occurred despite the fact that key stakeholders, Penan hunter-gatherers, "defended their claims to land in the most profoundly moral terms" (38-40).

Moreover, the findings of this study inherently question the viability of attitude surveys. There is no substitute for the statistical representativeness of survey methods. But surveys can only offer stakeholders discrete declarative statements about values and preferred management options, to which the respondent then agrees or disagrees. However, it is abundantly clear that values held by lay stakeholders are typically articulated discursively; they are embedded in the contextually, emotively, and morally rich stories and conversations through which we define ourselves and our actions in relation to natural systems. The ability for stakeholders to articulate the non-utilitarian qualities and values that best express why nature or specific logging practices matter is inhibited by the poverty of opportunities for expression that surveys impose. This problem is a significant one because an absence of opportunities for elaborated discussions of environmental values serves to either misrepresent the public or to relegate discussion to elite and, thus, non-democratic venues.

My point is not that all policy studies should be field studies. Most policy studies are confined by time – the anthropologist's inclination towards in-depth fieldwork or "deep hanging out"[10] over sustained periods is not often an option. It is therefore necessary to develop value-elicitation opportunities, frames, or contexts that resist the tendency to fit the articulation of values into economistic expressions, that seek alternatives to direct question-answer formats, and/or that denude value expressions of relevant moral or affective content. I have explored this possibility elsewhere (Satterfield 2001;

Satterfield, Slovic, and Gregory 2000). But let me state briefly that narrative-centred elicitation devices can be developed to improve the diversity of values expressed by participants in open-ended elicitation contexts.[11] Narrative passages of the type common throughout this book (think, for instance, of Steve Fuller's long passage about the ability of loggers to produce something physical or Andrew Simon's rapturous musings on science and becoming place-attached to Sumner Mountain), photographs, and descriptions of real-life environmental conflicts can be used to help stakeholders articulate their own narratives of place, meaning, and value. Overall, I have found that stakeholders (and study participants generally) are delightfully articulate about a wide variety of non-cost and non-utilitarian values when value elicitation exercises encompass rather than avoid affective, imagistic, storied, or morally meaningful content.

Pinning Down Stakeholder Values

Finally, it may seem nearly impossible to reconcile the central spirit of this book – the messy, creative, and culture-changing dialogue between activists situated amidst more powerful social forces – with the overwhelming demand from policy analysts for neat systematized summations of stakeholder values. Emerging of late, however, are several value-elicitation methods that mimic, indeed celebrate, the fact that values are difficult to pin down and are meaningful only when articulated within specific contexts. These are generally referred to as constructive processes.[12] They recognize (1) that value elicitation processes are very sensitive to the wording of questions and (2) that values are context-specific and not stable essential qualities that can be extracted from the stakeholder's mind and applied universally. Consequently, the context in which value information is being sought must be carefully specified using, in some cases, narrative and morally meaningful discourses as opposed to techno-rationalist ones (Satterfield, Slovic, and Gregory 2000). Discussion can follow a course that probes deeply and renders stakeholder thinking (or practice) visible, allowing participants to evaluate decisions about land management in a manner that closely reflects that thinking. The question about what constitutes an optimal balance between constructive processes (which are somewhat structured in that they help people think through decision-making problems) and those that provide ample room for the expression of values with which policy advisors have difficulty (e.g., those based on affective investment, enchantment, ethical propositions, or impassioned stories about nature and meaning) is an open one. Ultimately, it is a question fully appropriate both to anthropologists who have studied the implications of power, emotion, narrative expression, and the meaning of place in a global world and, equally, to the study of conflict as identities emerge and thicken by virtue of their engagement in impassioned dialogues about cultural futures and the land upon which they depend.

Notes

Chapter 1: Introduction

1   I often found myself equally moved by (1) the trauma of job loss and unwanted community change experienced by loggers and (2) the commitment to the common good of wild and biologically healthy forests on the part of environmentalists. In Chapter 3 I spell out more fully the implications of this predicament for fieldwork.

2   Caterpillar makes much of the heavy equipment used to build roads for logging.

3   The dispute is frequently reminiscent of Keith Thomas's (1983) history of the evolution of English mentality and its relationship to nature. He focuses on the period between 1500 and 1800, wherein "some long established dogmas about man's [sic] place in nature were discarded." At one time "uncultivated land meant uncultivated men," while later "[man's] right to exploit species for his own advantage was sharply challenged" (15).

4   Bourdieu (1977, 1985) addresses this kind of "awareness" when discussing his concepts of field and practice. He notes that individuals act with awareness of their social position and adjust their behaviour according to their perception of the social influence of their actions.

Chapter 2: The Cycle of History

1   Cronon (1996, 73) associates this sentiment with romantic notions of the sublime – "of those rare places on earth where one had more chance than elsewhere to glimpse the face of God."

2   Jerry Franklin of Corvallis Oregon's US Forest Experiment Station is appropriately recognized as one of very few professional foresters working to understand the ecological function of old-growth forests (Hays 1998, 177-79).

3   In biological terms, younger trees add biomass at a rapid rate, while the greater biomass of old trees means that most of their production goes into maintenance (Booth 1994, 25).

4   The premise that nature has intrinsic value is central to a new breed of environmental philosopher working to counter an excessively economistic view of nature (Callicott 1984; Rolston 1994).

5   I verified this fact by calling the number on the poster.

6   The European appellation for the area's Aboriginal population is actually the Aboriginal word for tall grasses – the most striking feature of the Willamette and Calapooia valleys (Boag 1992, 3).

7   By way of regional example, half of the Aboriginal population of the Puget Sound area had been wiped out by disease by 1840. Another great epidemic had yet to hit.

8   The gold rush also increased the pressure for both commodity goods and land as, in the early 1850s, prospectors moved into southern Oregon (Robbins 1997, 85).

9   This did not, however, discourage land speculators from luring naive easterners west on the suggestion that forest abundance equalled agricultural abundance.

10 This is thematically and technically reminiscent of attitudes surrounding recent log-scaling trials in Oregon (Painter 1992). The defendants are accused of removing more timber from public land than is "scaled" (i.e., reported). Environmentalists applaud the long overdue enforcement of the law, while some timber-dependent individuals anonymously protest

the sudden criticism of long-held (and therefore acceptable) patterns of behaviour (Hayden 1993). See also Roberts (1997, 46) for an account of US Forest Service efforts to abort investigations into timber theft.

11 Today selective cutting in roadless areas is conducted with either helicopters or horses.

12 Chokers occasionally broke, as they still do today, hurling downhill immense logs that annihilated everyone and everything in their path. A suddenly broken whiplashing cable could also sever whatever or whomever it contacted.

13 By "dormant fire" I mean the fuel load created by dead woody debris left on the ground after the cutting of timber.

14 Many attribute the roots of modern environmentalism to the conservation-efficiency movement; however, to the extent that they blend an ethic based on managerial and technical improvements to the use of physical resources with an environmental ethics based on the quality of land, marine, and forest systems, such attributions are false. These movements, as will become clear in subsequent pages, are unique and have "marked differences in sources of support" (Hays 1998, 380).

15 Apparently, Marsh's experiences in the rapidly exhausted Vermont woods motivated some of his thinking (Caufield 1990, 51).

16 Muir regarded transcendentalism as the "essential philosophy for interpreting the value of wilderness." A mentor (Jeanne Carr) from his undergraduate years at the University of Wisconsin introduced Muir to Emerson (Nash 2001, 125-6). Shabecoff (1993) also notes that Muir claimed to be influenced by Marsh.

17 That Yosemite and other national park land was pristine land free form human impact is a foundational myth central to the history of American environmentalism. This "uninhabited" land was, more accurately, a product of the dramatic decline of Native populations during the colonial period and of a reservation system that confined Aboriginal populations to vastly reduced territories (Cronon, Miles, and Gitlin 1992; Nash 2001).

18 This estimate includes all of the region's state and federally owned land.

19 A 1994 national survey of registered voters found that 74 percent of the population held a favourable impression of the Forest Service, while only 33 percent held a favourable impression of the Bureau of Land Management (Palmer 1995). Meanwhile, several studies have charged the Bureau of Land Management and the Forest Service with colluding with the timber industry (e.g., Taylor 1994).

20 Until the late 1970s, much of Boise-Cascade (a "giant" in its own right) was actually owned by Weyerhaeuser (Dunn 1977).

21 These details are summarized following the above referenced Oregon Office of the Secretary of State's *Blue Book*. See also Porter (1996, 75-76).

22 Walls (1996) locates the origins of the Paul Bunyan tales in the oral stories of dramatic deeds of nineteenth-century woods workers. Bunyan is the "fictitious giant lumberjack capable of doing the impossible all on his own" (125).

23 The Sierra Club, suspicious of the Forest Service's timber agenda and disappointed with its lack of attention to wilderness preservation, did not support the act (Booth 1994, 144).

24 This is the source of the quote "to keep every cog and wheel is the first precaution of intelligent thinking," which often appears on posters.

25 The Fish and Wildlife Service (FWS), along with the Bureau of Land Management, is located within the Department of the Interior. Employed within FWS are many of the scientists (biologists, hydrologists, ecologists, etc.) who oversee species and habitat protection. Because FWS is responsible for the listing of threatened or endangered species, it is at times forced to compete (intra-departmentally) with the Bureau of Land Management and (extra-departmentally) with the Department of Agriculture's Forest Service over timber sales. The FWS's political clout has grown in conjunction with the above noted public support for environmental concerns, a fact that has escalated the potential for conflict between these government agencies. To complicate matters, conflicts about forest priorities exist within various departmental divisions. Rank and file Forest Service employees often find themselves at odds with the more powerful "old guard" within that agency. During an informal conversation, a Forest Service employee told me that the agency "peons'" term for the "upper brass" is MFWIC. He had a button with these initials inside a red circle and slash. I was told that MFWIC stands for "Mother Fucker What's In Charge." Hirt (1994) does a superb job of detailing the revolt of conscience within the Forest Service since 1989 (see especially 271-92).

26 This happened in 1989, 1990, and (more thoroughly) in 1991.
27 For competing perspectives and definitions of employment impact, see Whitelaw 1995; Freudenberg, Wilson, and O'Leary 1998; Carroll, Daniels, and Kusel 2000; Carroll, Blatner, et al. 2000.
28 Under the new plan, the probability of protecting keystone species was, by many scientific standards, not sufficiently high; consequently, either "exempting" legislation or an agreement from environmental plaintiffs to drop their lawsuits was needed. Clinton gained the support of environmentalists by threatening to shield the plan from judicial review – an action that would have eliminated the courts as a key means for resisting unlawful timber sales (Hirt 1994, 291-92).

Chapter 3: Disturbances in the Field

1 Much more will be said about the identity-establishing processes in each of the subsequent chapters.
2 A fine-grained discussion of the distinction between Wise Use and the forest community movement can be found in Porter (1996, 146-54).
3 The OFCC's large membership partially reflects one of its fund-raising strategies. Members of resource-dependent communities are asked to donate one dollar per month to join, a fee virtually everyone can afford.
4 Individual groups within the coalition have their own budgets as well as paid and volunteer staff members.
5 For a complete history of the Millennium Grove dispute, see Porter (1996, 13-35). Porter cites several negotiating efforts between activists and Willamette Industries, who purchased the timber from the Forest Service. The industry's motives to benefit from the sale were always clear; the Forest Service, however, was duplicitous in its actions and communications with environmentalists.
6 "Tree sitting" involves taking up protective residence on a platform high in a tree that has been designated for cutting.
7 Their 1993 operating budget was $700,000. Individual membership at the time was $35.00, group membership was $100.00 (Stiak 1992; Pittman 1993).
8 Forty-four persons were interviewed once, and six were interviewed twice, for a total of fifty interviews. Thirty-two of these interviews are quoted at some length in subsequent chapters. I worked only with the local grassroots activists who were explicitly concerned with the spotted owl controversy. Eighteen of my interviews (six women and twelve men) were conducted with members of Oregon's local environmental community, while twenty-six interviews (twelve women and fourteen men) were conducted with members of pro-timber groups in two of Oregon's many timber-dependent communities. As also noted on page x, I oversampled timber communities in order to compensate for the fact that this population was less familiar to me than was that of the environmental community.

    Initially I thought I might have trouble finding women to interview within the world of loggers. As it turned out, loggers are absorbed for long hours every day in their work in the woods. This means that women have been left in charge of much of the public, pro-timber work. Among grassroots timber advocates, women staff the offices, prepare and often run the meetings, write and issue press releases, review contested timber sales, and so on. Among environmentalists, it was, surprisingly, much harder to find an equal number of similarly positioned (i.e., fully absorbed and active) men and women to interview due to the preponderance of men. During my research period, only one of the coalition groups under the AFGA umbrella was led by a woman; the executive director (and official head) of a second group was male, but its communications director was female.

    A second surprise was that timber advocates were much more willing to accommodate me in terms of time, hospitality, and so forth than were environmentalists. Notwithstanding initial suspicion or defensiveness towards an outsider such as myself, a phone call would often open any door in the small timber towns where I conducted my interviews. Conversely, when I attempted to develop contacts among environmentalists, with few exceptions only a formal letter with a detailed explanation of my goals opened the necessary doors. Ultimately, members of both groups appeared to enjoy the interview process, and the open-ended nature of my questioning was appreciated in a world where a ten-second, media-driven sound bite can mean everything.
9 Those speaking were county and state political representatives, grassroots and industry spokespersons, and one representative from the local chapter of the Audubon Society.

10 En route to the logging site I was complimented on my ability to drive aggressively, as I followed the bus on the winding roads.

11 On the way up in the car, Jim told me that the forestry student's incomplete degree was the result of laziness, while the math teacher was said to be too shy to get up in front of a classroom.

12 Jim Stratton was my source for logging terms.

13 The water runoff from logging uphill created the swamp-like conditions hospitable to alder.

14 The sale attracted the attention of activists because it would interrupt a wildlife corridor thought to join unique biological regions in the state's northern and southern territories. Criticism from state legislators and the faculty at Oregon State University as well as tree spiking and countless sit-ins and arrests failed to stop the cut.

15 This discrediting nickname for Forest Service employees is a play on "Freddie" Krueger of horror-movie ("slasher flick") fame.

16 The non-violent, direct-action workshop mentioned earlier was cancelled as most of the people present had already been through the training.

17 Caulk boots are spiked work boots that are ideal for walking the length of fallen trees or working one's way up slippery hillsides.

18 The existing divisions between environmentalists and wood workers unions worked against both groups with regard to their wish to gain support from the Clinton administration. Though labour had not been especially active at the grassroots level, as noted in the last chapter, union leaders were often invited to speak at conferences.

19 Andrea is one of the few prominent women within the AFGA coalition.

20 For an insightful study of the link between timber-community interests and the anti-gay movement, see Stein (2001), particularly her chapter, "Resentment's Roots."

Chapter 4: Negotiating Agency

1 Many environmental activists would argue that, given the vast tracts of forests lost forever, they have ended up with the smaller piece of the pie. Many of the land-use records confirm this point; nonetheless, during this study period, there is no denying that the Clinton Plan was an unprecedented victory – one that would have been inconceivable in the years prior.

2 Political advantages are rarely stable. But I am concerned primarily with this early period (1992-94), when the struggle for grassroots status was significant for both parties.

3 The ancestral phrase is taken from a talk given by Waite, who is quoted at length in the forthcoming pages. The segregationist analogy was often quoted in the media and also surfaced in interviews with Waite.

4 Caledon is a timber-dependent town located about 160 kilometres southwest of Portland, Oregon.

5 Yellow ribbons, derived from the plastic tape used for surveying a timber sale, have become a symbol for pro-timber sentiment. Most people date the association back to the 1987 Silver Fire Round-up in southern Oregon – a gathering that protested the halting of salvage logging at the site of the Silver Fire.

6 Taylor's integration of this construct with contemporary ideals of authenticity and individuality elaborates the politics of recognition more fully than is appropriate here.

7 The idea reflects the work of G.H. Mead (1934) and M. Bahktin (1981).

Chapter 5: Voodoo Science and Common Sense

1 The comments are taken from Clinton's closing remarks at the 1993 "Forest Summit" held in Portland, Oregon. The latter two of Clinton's three points are redundant in that, in order to be ecologically credible, one must draw from the latest scientific findings in forest, stream, and wildlife ecology. Obeying the Endangered Species Act (ESA), the National Forest Management Act (NFMA), and/or the National Environmental Policy Act (NEPA) involves considerable deference to guidelines, indicators, and practices set forth by the scientific community. (NFMA and NEPA also require a certain amount of public consultation.)

2 A shared, or figured, world is analogous to Geertz's "webs of meaning" – the context for meaning and interpretation, the imagined world that we assume to be true and, therefore, that informs our behaviour (Holland et al. 1998).

3 It is telling that AFGA activist Paul Wilson cited as one of his "best moments" the point at which ex-chief of the US Forest Service and wildlife biologist Jack Ward Thomas said to him: "Who wrote this [referring to a Forest Service appeal to suspend a timber sale]?

Somebody who really knew what they were doing wrote this." Wilson, who'd written the appeal, recalled thinking to himself: "This is not going to get any better. This is really good. I mean we had different [i.e., non-specialist] words, but we were beginning [to get the scientific picture]."

4 Both quotes in this paragraph are excerpted from my field notes.

5 I speak generally here and am not pretending to deny the existence, in both the environmental community and in timber communities, of people who maintain an adamant and wholehearted faith in science. There are also, of course, many sciences. I am examining reference by activists to any science; however, exploration of lay evaluations of specific disciplines would also be revealing.

6 See, especially, Leopold (1966).

7 This definition of enchantment is borrowed from *The New Shorter Oxford English Dictionary* (1993).

8 Leo Marx's (1964, 3) classic, *The Machine in the Garden*, documents the desire of settler Americans to tame and manicure the wild: "The pastoral ideal has been used to define the meaning of America ever since the age of discovery."

9 Stephen Tyler (1987) has developed some interesting ideas about the priority of "feelings" versus the "visual" in the human sensorium. Our linguistic system tends to equate thought, knowing, and seeing. To Tyler, science and common sense are joint products of this equation, not dichotomous phenomena. Dravidian language systems, in contrast, pair thought, knowledge, and feeling: to feel is to know.

10 For a review of "common sense" references, I recommend a perusing of the Web sites that pop up under "common sense," via www.google.com. Thomas Paine's, William James's and Einstein's definitions can be found there as well.

11 Interstate 5 is the north-south route through the state's most populated urban centres.

12 It is difficult for me to believe that Beverly is as restricted an Oregon traveller as she pretends, especially since the home and business she shares with her husband are located in a small town on a secondary highway. I suspect her self-casting is a way of being polite to me. In her telling we are both naive outsiders; I am not left to stand alone.

13 Holland et al. (1998, 58) note that several of Bourdieu's key concepts – namely, "field" and "habitus" – are too potent and pervasive a theme in his thought to be precisely delimited. I do, however, think that White's use aptly captures the spirit of the construct "habitus," to the extent that habitus reflects the internalization of social knowledge. Bourdieu (1977), however, also uses "habitus" to refer to the processes by which dominant culture is reproduced and unknowingly taken up by subordinate groups as the natural order of things.

14 I have not yet broached head-on the subject of local knowledges, which is that branch of science and technology studies concerned with empirical knowledge systems also known as "traditional ecological knowledge" and, sometimes, non-Western science (Nader 1996). Such studies emphasize situated learning and the importance of practical knowledge (Palsson and Helgason 1998). Sillitoe (1998b, 204) offers a broader definition: "local knowledge ... may relate to any knowledge held collectively by a population, informing interpretation of the world." The influence of these ideas is evident in this discussion, though I have chosen to focus on the intersection of ideas about labour, class, and experiential knowledge.

Chapter 6: Theorizing Culture

1 Conklin and Graham (1995, 695) refer to the "middle ground" thesis developed by historian Richard White, who uses the idea to describe the processes of confrontation, collaboration, negotiation, and innovation particular to Aboriginal/White relations in North America's Great Lakes region in the seventeenth to nineteenth centuries. Clifford (1997) employs a similar idea when writing about museums as "contact zones" encapsulating White/First Nation relations.

2 Several of these questions were inspired by Clifford (1997). In the last question, I refer to "major force," by which I mean those present in the thick of the day-to-day political machinations of Oregon's old-growth dispute. Many First Nations activists have, however, spoken and written on the subject more broadly and have ardently criticized both environmentalists and anthropologists for unnecessarily and naively romanticizing the relationship between Aboriginal peoples and nature (Biolsi and Zimmerman 1997).

3 I am making only a general point about the Enlightenment period. There were, in actuality, many Enlightenments; debates shifted over time and across European borders.

4 Rousseau is, as noted, popularly associated with Aboriginal nobility claims – an affiliation in the public, and even academic, mind that is not upheld by the scholarly literature. Early on, Fairchild (1955) argued that there is no compelling evidence to prove that Rousseau either seriously believed in, or invented the idea of, Aboriginal nobility, or "noble savagism."

5 In Wolf's (1982) familiar image, cultural relativism was imagined by Boas and his students as a set of billiard balls on the table, each banging against one another but each operating as distinct, self-contained wholes (or even impermeable units).

6 A recent survey of trends in anthropological scholarship has upheld this claim. Gupta and Ferguson (1997, 13) conclude that prestige among one's disciplinary colleagues is still defined by "grasshut" anthropology (i.e., working in "foreign" and "exotic" fieldwork locations). Shankman and Ehlers (2000) examine article submissions in major anthropology journals, only to find that regional scholarship in anthropology has shifted from an early-century emphasis on North American fieldwork sites to a late-century emphasis on Oceania, Asia, and Africa. They interpret this shift as an expression of career opportunities that favour "exotic" and "pure" research in contrast to "domestic" and applied research.

7 Redford first published the article in the spring 1990 edition of *Orion Nature Quarterly*.

8 Not all concerned, including Redford, recognize this right. Redford (1991, 46) criticizes an Aboriginal spokesperson at the 1981 International NGO Conference on Indigenous People for claiming: "In the world of today there are two systems, two different irreconcilable 'ways of life.' The Indian world – collective, communal, human, respectful of nature, and wise – and the western world – greedy, destructive, individualist, and enemy of nature."

9 Buege (1996, 74) speaks of a direct connection to the environment as a form of ecological determinism: "the people who are ecologically noble would not be so in another environment." I am less convinced that this is the assumption behind the stereotype. It may be, rather, that we assume that nobility is intrinsic to the person and travels with him or her (i.e., a form of essentialism). Or we assume that Aboriginal peoples do not or should not travel beyond their Aboriginal home as that would dilute their authenticity.

10 Interestingly, nostalgia is a synthesis of the Greek terms "nostos" (to return home) and "algia" (a painful condition) (Davis 1979, 1).

11 As Richard White (1996) has pointed out, it is as though Aboriginal North America was once a recreational (not a working) paradise, and even Lewis and Clark's westward journey was not regarded as hard physical work (which included numerous encounters with Aboriginal peoples working the land) but, rather, one long backpacking journey across the landscape.

12 Nancy Langston's (1995, 6) *Forest Dreams and Forest Nightmares* offers a fascinating discussion of these points. She notes that "there is no single moment in the past we can point to and say: 'That's what forests were really like before people started messing around with them.' Nonetheless, Forest Service policy has mandated that the inter-mountain west's Blue Mountain forests, about which she writes, be managed to resemble the forests that were there before Whites arrived.

13 The term "symbolic capital" stems from Bourdieu's (1977) supposition that the words and actions of high-status persons are granted more attention and credit than are the same words and actions delivered by a low-status person.

14 I do not mean to discount the fact that there is something we can legitimately call authentic attachment to place; anthropologists have offered some very elegant writing on the subject (e.g., Basso 1996). Nor do I mean to suggest that loggers' and environmentalists' attachments to places are not deeply felt and passionately defended. I do, however, mean to examine why claims to place matter so profoundly in this dispute, how such claims are made, and the problems that result from them.

15 Authenticity, as expressed for White audiences, is also indexed by traditional dress and the continuity of cultural practices – points I will address shortly.

16 Waite, like Dawson, uses the term "lifestyle."

17 Let me be absolutely clear: I do not mean to suggest that environmental movements consist entirely of elite or simply middle-class people. While this was often the case in the Oregon ancient forests movement at the time of this study, several of these activists did and do consciously eschew the possible material benefits of their educational training as part of their larger effort to bring forth a world that does not exhaust its natural resources. Moreover, over the last decade there is ample evidence of a synthesis of anti-poverty, anti-racism movements, and environmental movements – a phenomenon known

more broadly as the environmental justice movement (Bullard 1990; Johnston 1994, 1997; Szasz 1994).

18 For more elaborate theoretical explanations of this human actor/tool-wielding portrait of social process and change, see Holland et al.'s (1998) chapter entitled "Figured Worlds."

19 Interestingly, the claim that environmentalists are misanthropic has become the focus of attention for anthropologists who point to the phenomenon of "green imperialism," or "green colonialism" (wherein First World activists protect rare and biologically rich territory at the expense of Aboriginal rights). Others consider the criticism overstated and point to the growing awareness among green activists of the plight of Aboriginal peoples that have been dislocated at the expense of nature reserves.

20 This argument has also been effectively made by historian Richard White (1996) in his provocative essay, "Are You an Environmentalist or Do You Work for a Living?"

21 Evans completed two years of college on the GI Bill following the Vietnam War. He suspended his studies for economic reasons after the first of his children was born.

22 Similarly uncensored passages from environmentalists were recorded above. Admittedly, determining whether to rid ethnographic reporting of material that might cast an unfavourable light on those it represents is profoundly difficult. Politically sensitive (and, in this case, avowedly prejudiced) material, as Bourgois (1995) has observed, leaves the author torn between casting a "bad light" on those represented and being complicit with those who would dismiss political dissent as irrational noise from socially marginalized people. The point is neither to excuse nor to ignore the behaviour but, rather, to understand it as part of the larger socio-political circumstances within which it arises. An examination of this conundrum can be found in Bourgois's superb ethnographic study of poverty and the crack-cocaine trade (see, especially, 11-19).

23 I do not mean to suggest that the tug-of-war between loggers and environmentalists over cultural legitimacy "causes" racism in activists, especially timber-dependent activists. Racism is ubiquitous and undoubtedly exists on both sides, though it may express itself differently. I do mean to say, however, that the competition over resources that is the result of federal timber policies that favour corporate access and short-term economic gain and/ or disallow the rights of pre-White settlement populations generally fosters aggressive competition between those left to battle over the fate of remaining forests.

24 For a good analysis of the class-based denigration of loggers, see Foster (1993) and Brown (1995). That nature is regarded by many as sacred should be obvious, but discussions on this point can be found in Milton (1999), Dunlap and Scarce (1991), and Satterfield (2001).

25 These acts are specified in some detail in Chapter 2.

26 The suggestion that a society or ethnic group can be quantified as having a greater or lesser amount of culture is, like the noble savage premise, strongly resisted by contemporary anthropologists. It suggests a hierarchical ordering of cultural practices from better to worse, more valid to less valid, and thus defeats the strongly held belief in cultural pluralism noted above. But again, this is not to say that the broader public abides by these academic sensibilities. The point here is to examine what is said rather than to wish that expressions of popular culture obeyed disciplinary rules.

Chapter 7: Irrational Actors

1 For the curious, the following offer good reviews of the anthropology and sociology of emotions (Bessnier 1990; Kusserow 1999; Planalp 1999; White 1992).

2 The rules for emotional control are cross-culturally variable. Problems of reductionism aside, southern Italians are widely recognized for their emotional extroversion, whereas Briggs's (1970) influential study of the Utku, an Inuit community in the Canadian Arctic, demonstrates the centrality of affective restraint to a social life characterized by confined winter quarters, harsh physical conditions, and an ethos of interdependency and reciprocity. Wikan (1990) illuminates a similar principle in the Balinese context; she artfully conveys the nuances of emotional labour that must be invested as people calibrate their feelings in order to conform to the strict emotive-behavioural norms of Balinese society.

3 William James (1901) is often quoted as asking his students to imagine a person entirely devoid of emotion, an inquiry meant to conjure up a listless and lifeless being.

4 The first reference here is to Catherine Lutz's (1988) chapter on "Emotion, Thought, and Estrangement" in *Unnatural Emotions: Everyday Sentiments on a Micronesian Atoll and Their*

Challenge to Western Theory. A version of this chapter was initially published in 1986 as "Emotion, Thought, and Estrangement: Emotion as a Cultural Category" in the journal *Cultural Anthropology.*

5 The notion of authoring selves is detailed more fully in Holland et al. (1998).

6 Notice, too, that the "personal stake" of timber activists is then portrayed as the result of "real" problems; that is, economic woes. Imparted here is the idea that legitimate concerns are confined to those of human livelihood (see Gary's later comments – lines 23 to 24), a position meant to delegitimize concern with the "lesser," subordinate world of nature. This salience, in emotion talk, of the moral appropriateness of concern for nature versus concern for human communities will be examined more fully in the later portions of this chapter.

7 The proposition that women are better suited, or innately inclined, to care for the human and non-human worlds has, appropriately, been challenged by partnership models of ethical practice. "A partnership ethic treats humans (including male partners and female partners) as equals in personal, household, and political relations and humans as equal partners with (rather than controlled by or dominant over) nonhuman nature" (Merchant 1992, 188). Moreover, a partnership model permits feelings of intimacy and compassion between sexually, racially, and economically different humans as well as between humans and non-humans.

8 The female activist who resisted the ecofeminist perspective did so emphatically by banging her fist on the table and insisting that she and all women should not bear the moral burden of care towards the non-human world.

9 Chapter 6 made similar reference to strategic essentialism in its discussion of invocations of ecological nobility; there, also, the insights of previously dismissed or disparaged members of society were actively promoted.

10 I did not interview Richardson as he was not based in Oregon; I quote solely from speaking engagements that took place in Oregon.

11 Michelle Rosaldo (1980) has similarly argued, in her much cited *Knowledge and Passion: Illongot Notions of Self and Social Life,* that talk about emotions is encoded (like symbols) with information about a culture member's understanding and moral organization of their social world.

12 As with Proctor (1996), I must offer the caveat that the inner-group unity implied by this two-prong distinction is overly simplistic. There are many divergent interests within both camps.

13 Many definitions of the self have been offered, though Murphy's phrase: "the individual person as the object of [her/his] own perception" is especially succinct (quoted in Bock 1995, 197). To say that all human cultures express a self-concept is not to say that every language possesses a term that can be glossed in English as "self." For a discussion of this point, see Parish (1994).

14 I tend to agree with Markus and Kitayama (1991), who argue that all societies have both types of self-orientations but that one type tends to dominate or be marked as superior. I also agree with Holland and Kipnis (1994), who argue that some emotions (such as embarrassment) work dialectically with a sociocentric conception of the self, whereas others are more suited to an egocentric conception. For a recent and thorough treatment of the central problems and inconsistencies inherent in theories of the self and the "West/Other" divide, see Kusserow (1999).

15 Abram's (1996) relational epistemology, following the work of Merleau-Ponty, considers the perceptual implications of the Gaia hypothesis. He argues that the natural world is not inanimate, awaiting whatever we project onto it, but, rather, animate, living, breathing, and thus capable, on many levels, of being an active participant in the intersubjective field that is perception. Milton (2000) has similarly noted that personhood as it applies to nature is rooted in perceptual experience.

16 This comment was provided by the editor of a small pro-timber quarterly published by the Oregon Forest Community Coalition.

17 It is worth noting that, after Richardson's ecocentric talk (and consistent with environmentalists' disinterest in emotional control), he was granted a sustained and standing ovation. Milling about in the conference halls, I could hear people praising the talk and asking others if they'd heard it. Interestingly, I noticed that audience members were often unable to articulate how or why it moved them, only that it did.

Chapter 8: A Concluding Discussion

1 For a recent review of this literature, see Little (1999).

2 Sheridan's (1988) probing and thoughtful ethnography of rancher-farmers in northwest Mexico does reveal the tendency for subjugated peoples to fight with one another as well as against the inequities inherent in the region's pattern of land and water distribution.

3 To make myself clear: I do not mean to suggest that grassroots loggers and environmentalists share, in this dispute, the *same* position. But both are marginalized parties – though differently so. Loggers, most obviously, are marginalized due to class and/or their activism. Environmentalists benefit significantly from social capital bequeathed by their largely middle-class status, but they are also marginalized as activists and as individuals who consciously choose to forego the usual economic power inherent in middle-class life.

4 I mean to limit my point about the triangular shape of cultural production to studies of environmental conflict in anthropology and sociology. Social theorists have addressed, within other contexts, an over-emphasis on resistance struggles that lead to an avoidance of how competing subordinate groups affect one another. Bourdieu's (1988) *Homo Academicus*, for instance, emphasizes competition among groups, whereas Ginsberg's (1989) outstanding study of the abortion debate looked carefully at both pro-choice and pro-life positions.

5 One can also ask how the more powerful appropriate (i.e., co-opt) discourse or imaginaries from below and to what end – corporate or otherwise?

6 In a survey (Satterfield and Gregory 1998, 632), of Ontario, Canada's general public, residents of timber-dependent communities, and, specifically, timber-dependent households, 87 percent to 91 percent of respondents agreed with the species egalitarian ideal. Between 68 percent and 73 percent agreed with the statement about being attracted to nature's "spiritual qualities." With regard to both questions, support was higher among members of timber-dependent households and communities (as opposed to the "general" public). Just what such endorsements mean at the level of acceptable land management policies is, of course, a different matter and is taken up more fully in the paper.

7 Ortner's "borderlands" reference is taken from Rosaldo (1989, 1994), whose *Culture and Truth*, along with her subsequent "Race and Other Inequalities," looked at the interplay of power, gender, and race along the United States-Mexico border.

8 The single best example I can think of comes not from environmental studies but from Bourgois's (1995) study of street culture amidst the crack trade in East Harlem. He states that "although street culture emerges out of a personal search for dignity and a rejection of racism and subjugation, it ultimately becomes an active agent in personal degradation and community ruin" (9).

9 Interestingly, the term "stakeholder" comes from the business community and, in this sense, is indicative of the prominence of market-driven approaches to public participation.

10 This phrase is borrowed from Rosaldo and is quoted in Clifford (1997, 56).

11 Ethnographic interviews of the kind used here and elsewhere (see, especially, Kempton, Boster, and Hartley 1995) are also a widely used and productive option.

12 See especially Gregory, Lichtenstein, and Slovic (1993); Satterfield and Gregory (1998).

References

Abram, D. 1996. *The Spell of the Sensuous*. New York: Vintage.

Alvard, M. 1993. "Testing the 'Ecologically Noble Savage' Hypothesis: Interspecific Prey Choice by Piso Hunters of the Amazonian Peru." *Human Ecology* 21(4): 355-87.

Anderson, E. 2000. "Is It All Politics? Interaction and Event in Human Ecology." Paper presented at the 99th Annual Meeting of the American Association of Anthropology, San Francisco, CA, November.

Armstrong, S., and J. Botzler. 1993. *Environmental Ethics: Divergence and Convergence*. New York: McGraw-Hill.

Averill, J. 1994. "Emotion Unbecoming and Becoming." In P. Eckman and R. Davidson, eds., *The Nature of Emotions: Fundamental Questions*, 265-69. Oxford: Oxford University Press.

Bahktin, M.M. 1981. *The Dialogic Imagination: Four Essays by M.M. Bahktin*. Ed. M. Holquist. Austin, TX: University of Texas.

Barth, F. 1995. "Other Knowledge and Other Ways of Knowing." *Journal of Anthropological Research* 51: 65-68.

Basso, K.H. 1996. *Wisdom Sits in Places: Landscape and Language among the Western Apache*. Albuquerque: University of New Mexico.

Beck, U. 1992. *Risk Society: Towards a New Modernity*. London: Sage Publications.

Benedict, R. 1932. "Configurations of Culture in North America." *American Anthropologist* 34: 1-27.

Berglund, E. 1998. *Knowing Nature, Knowing Science*. Cambridge: White Horse Press.

Berkhofer, R.F. 1978. *The White Man's Indian: Images of the American Indian from Columbus to the Present*. New York: Alfred A. Knopf.

Bessnier, N. 1990. "Language and Affect." *Annual Review of Anthropology* 19: 419-51.

Biersack, A. 1999. "From the 'New Ecology' to the New Ecologies." *American Anthropologist* 101(1): 5-18.

Biolsi, T., and L. Zimmerman, eds. 1997. *Indians and Anthropologists: Vine Deloria Jr. and the Critique of Anthropology*. Tucson: University of Arizona.

Boag, P. 1992. *Environment and Experience: Settlement Culture in Nineteenth-century Oregon*. Berkeley: University of California.

Bock, P. 1988. "The Importance of Erving Goffman to Psychological Anthropology." *Ethos* 16(1): 3-20.

—. 1995. *Rethinking Psychological Anthropology: Continuities and Change in the Study of Human Action*. Prospect Heights, IL: Waveland Press.

Booth, D. 1994. *Valuing Nature: The Decline and Preservation of Old-Growth Forests*. Lanham, MD: Rowman and Littlefield Publishers.

Bourdieu, P. 1977. *Outline of a Theory of Practice*. Trans. R. Nice. New York: Cambridge University Press.

—. 1985. "The Genesis of the Concepts of 'Habitus' and 'Field.'" *Sociocriticism* 2(2): 11-24.

—. 1988. *Homo Academicus*. Cambridge: Polity Press.

—. 1990. *The Logic of Practice*. Trans. R. Nice. Stanford, CA: Stanford University Press.

—. 1993. *The Field of Cultural Productions: Essays on Art and Literature*. New York: Columbia University Press.

Bourgois, P. 1995. *In Search of Respect: Selling Crack in El Barrio*. New York: Cambridge University Press.

Brick, P. 1995. "Determined Opposition: The Wise Use Movement Challenges Environmentalism." *Environment* 37(8): 17-20, 36-42.

Brick, P., and R. McGreggor Cawley, eds. 1996. *A Wolf in the Garden: The Land Rights Movement and the New Environmental Debate*. Lanham, MD: Rowman and Littlefield Publishers.

Briggs, J. 1970. *Never in Anger*. Cambridge, MA: Harvard University Press.

Brosius, P. 1999a. "Analysis and Interventions: Anthropological Engagements with Environmentalism." *Current Anthropology* 40(3): 277-309.

—. 1999b. "Green Dots, Pink Hearts: Displacing Politics from the Malaysian Rain Forest." *American Anthropologist* 101(1): 36-57.

Brosius, P., A. Tsing, and C. Zerner. 1998. "Representing Communities: Histories and Politics of Community-based Natural Resource Management." *Society and Natural Resources* 11: 157-68.

Brown, B. 1995. *In Timber Country*. Philadelphia: Temple University Press.

Buege, D.J. 1996. "The Ecologically Noble Savage." *Environmental Ethics* 18: 71-88.

Bullard, R.D. 1990. *Dumping in Dixie: Race, Class, and Environmental Quality*. Boulder, CO: Westview Press.

Callicott, J.B. 1984. "Non-anthropocentric Value Theory and Environmental Ethics." *American Philosophical Quarterly* 21: 299-309.

—. 1995. "Environmental Ethics: Overview." In W.T. Reich, ed., *The Preservation of Species*, 138-72. Princeton, NJ: Princeton University.

Carney, J. 1996. "Converting the Wetlands, Engendering the Environment." In R. Peet and M. Watts, eds., *Liberation Ecologies: Environmental, Development, Social Movements*, 165-87. New York: Routledge.

Carroll, M.S. 1995. *Community and the Northwestern Logger: Continuities and Changes in the Era of the Spotted Owl*. Boulder, CO: Westview Press.

Carroll, M., K. Blatner, F. Alt, E. Schuster, and A. Findley. 2000. "Adaptation Strategies of Displaced Idaho Woods Workers: Results of a Longitudinal Panel Study." *Society and Natural Resources* 13: 95-113.

Carroll, M., S. Daniels, and J. Kusel. 2000. "Employment and Displacement among Northwestern Forest Products Workers." *Society and Natural Resources* 13: 151-56.

Carson, R. 1962. *Silent Spring*. New York: Houghton Mifflin.

Caufield, C. 1990. "The Ancient Forest." *The New Yorker*, 14 May, 46-84.

Cawley, R. McGreggor. 1993. *Federal Lands, Western Anger*. Lawrence: University of Kansas Press.

Clark, Norman. 1970. *Mill Town: A Social History of Everett*. Seattle: University of Washington Press.

Clifford, J. 1988. *The Predicament of Culture: Twentieth-century Ethnography, Literature, and Art*. Cambridge, MA: Harvard University.

—. 1997. *Routes: Travel and Translation in the Late Twentieth Century*. Cambridge, MA: Harvard University Press.

Clifford, J., and G. Marcus, eds. 1986. *Writing Culture*. Berkeley: University of California Press.

Conklin, B.A. 1997. "Body Paint, Feathers, and VCRs: Aesthetics and Authenticity in Amazonian Activism." *American Ethnologist* 24(4): 711-37.

Conklin, B.A., and L. Graham. 1995. "The Shifting Middle-Ground: Amazonian Indians and Eco-politics." *American Anthropologist* 97: 695-710.

Cooper, S.F. 1850. *Rural Hours*. New York: George P. Putnam.

Cronon, W. 1989. "Comments on Landscape, History, and Environmental Change." *Journal of Forest History* 33: 125.

—, ed. 1996. *Uncommon Ground: Rethinking the Human Place in Nature*. New York: W.W. Norton and Company.

Cronon, W., G. Miles, and J. Gitlin, eds. 1992. *Under an Open Sky: Rethinking America's Western Past*. New York: W.W. Norton.

Davis, F. 1979. *Yearning for Yesterday: A Sociology of Nostalgia*. New York: The Free Press.

della Porta, D., and M. Diani. 1999. *Social Movements*. Malden, MA: Blackwell Publishers.

Descola, P. 1996. "Constructing Natures: Symbolic Ecology and Social Practice." In P. Descola and G. Palsson, eds., *Nature and Society: Anthropological Perspectives*, 82-102. New York: Routledge.

Devall, B. 1992. "Deep Ecology and Radical Environmentalism." In R.E. Dunlap and A.G. Mertig, eds., *American Environmentalism: The U.S. Environmental Movement, 1970-1990*, 51-62. Washington, DC: Taylor and Francis.

—. 1993. *Clearcut*. San Francisco: Earth Island Press.

Diamond, J. 1992. *The Third Chimpanzee: The Evolution and Future of the Human Animal*. New York: Harper Collins.

Dietrich, W. 1992. *The Final Forest*. New York: Simon and Schuster.

Dove, M. 1996. "Process versus Product in Bornean Augury: A Traditional Knowledge System's Solution to the Problem of Knowing." In R. Ellen and K. Gukui, eds., *Redefining Nature*, 557-96. Oxford, UK: Berg.

Downey, G., and J.D. Rogers. 1995. "On the Politics of Theorizing in a Postmodern Academy." *American Anthropologist* 97(2): 269-81.

Dunlap, R., and R. Scarce. 1991. "The Polls–Poll Trends: Environmental Problems and Protection." *Public Opinion Quarterly* 55: 713-34.

Dunn, M.G. 1977. "Kinship and Class: A Study of the Weyerhaeuser Family." PhD diss., University of Oregon, Eugene.

Durbin, K. 1996. *Tree Huggers: Victory, Defeat and Renewal in the Northwest Ancient Forest Campaign*. Seattle: The Mountaineers.

Durbin, K., and R. Eisenbart. 1993. "Timber Timeline: Fight for the Forest." *Oregonian*, 25 March, A10.

Durbin, K., and P. Koberstein. 1990. "Special Report: Forests in Distress." *Oregonian*, 15 October, 1, 24-27.

Durkheim, E. 1965. *The Elementary Forms of Religious Life*. New York: Free Press.

Ellen, R., and K. Fukui, eds. 1996. *Redefining Nature*. Washington, DC: Berg.

Escobar, A. 1999. "After Nature: Steps to an Antiessentialist Political Ecology." *Current Anthropology* 40(1): 1-30.

Fairchild, H.N. 1955. *Noble Savage: A Study in Romantic Naturalism*. New York: Russell and Russell.

Forest Ecosystem Management Assessment Team (FEMAT). Team Leader. 1993. *Forest Ecosystem Management: An Ecological, Economic, and Social Assessment*. Washington, DC: GPO (FEMAT report).

Fortmann, L.P., J. Kusel, and S.K. Fairfax. 1989. "Community Stability: The Foresters' Fig Leaf." In D.C. Le Master and J.H. Beuter, eds., *Community Stability in Forest-based Economies*, 44-50. Portland, OR: Timber Press.

Foster, J.B. 1993. "The Limits of Environmentalism without Class: Lessons from the Ancient Forest Struggle of the Pacific Northwest." In *Capitalism, Nature, Socialism* (pamphlet), 3-34. New York: Monthly Review Press.

Foucault, M. 1978. *The History of Sexuality*. Trans. Robert Hurley. New York: Pantheon Books.

—. 1979. *Discipline and Punish: The Birth of the Prison*. Trans. Alan Sheridan. New York: Random House.

Franklin, J., and K. Kohm. 1999. "How Science Altered One View of the Forest." *Research in Social Problems and Public Policy* 7: 243-51.

Franklin, J., and R.H. Waring. 1980. "Distinctive Features of the Northwestern Coniferous Forest: Development, Structure, and Function." In R.H. Waring, ed., *Forests: Fresh Perspectives from Ecosystem Analysis. Proceedings of the 40th Annual Biology Colloquium*, 59-86. Corvallis: Oregon State University Press.

Franklin, S. 1995. "Science as Culture, Culture as Science." *Annual Review of Anthropology* 24: 163-84.

Freudenberg, W., L. Wilson, and D. O'Leary. 1998. "Forty Years of Spotted Owls? A Longitudinal Analysis of Logging Industry Job Losses." *Sociological Perspectives* 41(1): 1-26.

Geertz, C. 1973. *The Interpretation of Cultures*. New York: Basic Books.

—. 1983. *Local Knowledge: Further Essays in Interpretive Anthropology*. New York: Basic Books.

Gezon, L. 1999. "Of Shrimps and Spirit Possession: Toward a Political Ecology of Resource Management in Northern Madagascar." *American Anthropologist* 101(1): 58-67.

Ginsberg, F. 1989. *Contested Lives: The Abortion Debate in an American Community*. Berkeley: University of California Press.

Goffman, E. 1963. *Stigma: Notes on the Management of Spoiled Identities.* Englewood Cliffs, NJ: Prentice-Hall.

Goodenough, W. 1996. "Navigation in the Western Carolinas: A Traditional Science." In L. Nader, ed., *Naked Science: Anthropological Inquiries into Boundaries, Power and Knowledge,* 29-42. London: Routledge.

Gramsci, A. 1971. *Selections from the Prison Notebooks.* New York: International Publishers.

Grande, S.M. 1999. "Beyond the Ecologically Noble Savage: Deconstructing the White Man's Indian." *Environmental Ethics* 21: 307-20.

Gregory, R. 2000. "Using Stakeholder Values to Make Smarter Environmental Decisions." *Environment* 42(5): 34-44.

Gregory, R., S. Lichtenstein, and P. Slovic. 1993. "Valuing Environmental Resources: A Constructive Approach." *Journal of Risk and Uncertainty* 7: 177-97.

Grossman, L. 1998. *The Political Economy of Bananas.* Chapel Hill: University of North Carolina Press.

Gupta, A., and J. Ferguson. 1992. "Space, Identity, and the Politics of Difference." *Cultural Anthropology* 7(1) (theme issue).

—, eds. 1997. *Anthropological Locations: Boundaries and Grounds in Field Science.* Berkeley: University of California Press.

Hamilton, G. 2001. "Sensitized Loggers Learn to Deal with Protesters." *Vancouver Sun,* 15 January, A1, A6.

Harwood, J. 2002. "Bush Official Favors Increase in Logging." *Register Guard,* 13 January.

Hayden, C. 1993. "The Bill Jantzer Story: The Trials and Tribulations of a Local Logger." *Sneak Preview* 5 May, 8-9.

Hays, S. 1958. *Conservation and the Gospel of Efficiency.* Cambridge, MA: Harvard University Press.

—. 1998. *Explorations in Environmental History.* Pittsburgh: University of Pittsburgh Press.

Headland, T. 1997. "Revisionism in Ecological Anthropology." *Current Anthropology* 38(4): 605-15.

Hecht, S., and A. Cockburn. 1989. *The Fate of the Forest: Developers, Destroyers, and Defenders of the Amazon.* New York: Verso.

Hess, D. 1992. "Introduction: The New Ethnography and the Anthropology of Science and Technology." *Knowledge and Society: The Anthropology of Science and Technology* 9: 1-26.

—. 1995. *Science and Technology in a Multicultural World.* New York: Columbia University Press.

Hirt, P.W. 1994. *A Conspiracy of Optimism.* Lincoln: University of Nebraska Press.

Hochschild, A.R. 1979. "Emotion Work, Feeling Rules, and Social Structure." *American Journal of Sociology* 85: 551-75.

—. 1983. *The Managed Heart.* Berkeley: University of California Press.

Holland, D., and W. Kempton. 1999. "New Social Movements in Ethnographic Perspective: Forms of Environmental Activism in the United States." Paper presented at the annual meeting of the American Anthropological Association, November.

Holland, D., and A. Kipnis. 1994. "Metaphors for Embarrassment and Stories of Exposure: The Not-So-Egocentric Self in American Culture." *Ethos* 22(3): 316-42.

Holland, D., W. Lachicotte, Jr., D. Skinner, and C. Cain. 1998. *Identity and Agency in Cultural Worlds.* Cambridge, MA: Harvard University Press.

Holland, D., and J. Lave, eds. 2001. *History in Person: Enduring Struggles, Contentious Practice, Intimate Identities.* Santa Fe, NM: School of American Research Press.

Ignatieff, M. 2000. *The Rights Revolution.* Toronto: House of Anansi Press.

Irwin, A., and B. Wynne, eds. 1996. *Misunderstanding Science? The Public Reconstruction of Science and Technology.* Cambridge: Cambridge University Press.

James, W. 1901. *Principles of Psychology.* London: Macmillan.

James-Duguid, C. 1996. *Work as Art: Idaho Logging as an Aesthetic Moment.* Moscow, ID: University of Idaho Press.

Jasanoff, S., G. Murkle, I. Petersen, and T. Pinch, eds. 1995. *The Handbook of Science and Technology Studies.* Thousand Oaks, CA: Sage Publications.

Jasper, J. 1997. *The Art of Moral Protest.* Chicago: University of Chicago Press.

—. 1998. "The Emotions of Protest: Affective and Reactive Emotions in and around Social Movements." *Sociological Forum* 13: 397-424.

Jasper, J., and J. Paulson. 1995. "Recruiting Strangers and Friends: Moral Shocks and Social Networks in Animal Rights and Antinuclear Protest." *Social Problems* 42: 493-512.

Johnston, B.R., ed. 1994. *Who Pays the Price? The Sociocultural Context of Environmental Crisis*. Washington, DC: Island Press.

—. 1997. *Life and Death Matters: Human Rights and the Environment at the End of the Millennium*. Walnut Creek, CA: AltaMira Press.

Johnston, H., E. Laraña, and J. Gusfield. 1994. "Identities, Grievances, and New Social Movements." In E. Laraña, H. Johnston, and J. Gusfield, eds., *New Social Movements*, 3-35. Philadelphia: Temple University Press.

Jones, E., A. Farina, A. Hastorf, H. Markus, D. Miller, R. Scott, and R. de S. French. 1984. *Social Stigma: The Psychology of Marked Relationships*. New York: W.H. Freeman.

Joseph, S. 1993. "Fieldwork and Psychosocial Dynamics of Personhood." *Frontiers* 8(3): 9-32.

Kempton, W., J. Boster, and J. Hartley. 1995. *Environmental Values in American Culture*. Cambridge, MA: Massachusetts Institute of Technology.

Kluckhohn, F.R. 1950. "Dominant and Substitute Profiles of Cultural Orientations." *Social Forces* 28(4): 376-93.

Kraus, N., T. Malmfors, and P. Slovic. 1992. "Intuitive Toxicology: Expert and Lay Judgments of Chemical Risks." *Risk Analysis* 12: 215-32.

Kubler-Ross, E. 1969. *On Death and Dying*. New York: Macmillan.

Kunreuther, H., and P. Slovic. 1996. "Science, Values, and Risk." *Annals of the American Academy of Political and Social Science* 545: 116-25.

Kusserow, A. 1999. "Crossing the Great Divide: Anthropological Theories of the Western Self." *Journal of Anthropological Research* 55(4): 541-62.

Ladd, E.C., and K.H. Bowman. 1995. *Attitudes toward the Environment: Twenty-five Years after Earth Day*. Washington, DC: AEI Press.

Lange, J. 1996. "The Logic or Competing Information Campaigns: Conflicts over Old Growth and the Spotted Owl." In P. Brick and R. McGreggor Cawley, eds., *A Wolf in the Garden*, 135-50. Lanham, MD: Rowman and Littlefield Publishers.

Langston, N. 1995. *Forest Dreams and Forest Nightmares: The Paradox of Old Growth in the Inland West*. Seattle: University of Washington Press.

Laraña, E., H. Johnston, and J. Gusfield, eds. 1994. *New Social Movements*. Philadelphia: Temple University Press.

Lave, J., and E. Wegner. 1991. *Situated Learning: Legitimate Peripheral Participation*. Cambridge: Cambridge University Press.

Lee, R.G. 1994. *Broken Trust, Broken Land: Freeing Ourselves from the War over the Environment*. Wilsonville, OR: Bookpartners.

Leopold, A. 1966. *A Sand County Almanac: With Other Essays on Conservation from Round River*. New York: Oxford.

Levy, R., and M. Rosaldo, eds. 1983. "Issues Devoted to Self and Emotion." *Ethos* 11 (Special Issue).

Lincoln, B. 1989. *Discourse and the Construction of Society: Comparative Studies of Myth, Ritual, and Classification*. New York: Oxford University Press.

Little, P. 1999. "Environments and Environmentalism in Anthropological Research: Facing a New Millennium." *Annual Review of Anthropology* 28: 253-84.

Long, D. 1995. "Pipe Dreams: Hetch Hetchy, the Urban West, and the Hydraulic Society Revisited." *Journal of the West* (July): 19-31.

Lutz, C. 1986. "Emotion, Thought, and Estrangement: Emotion as a Cultural Category." *Cultural Anthropology* 1: 287-309.

—. 1988. *Unnatural Emotions*. Chicago: University of Chicago.

—. 1990. "Engendered Emotion: Gender, Power, and the Rhetoric of Emotional Control in American Discourse." In C. Lutz and L. Abu-Lughod, eds., *Language and the Politics of Emotion*, 69-91. New York: Cambridge University Press.

Marcus, G. 1995. "Ethnography in/of World Systems: The Emergence of Multi-sited Ethnography." *Annual Review of Anthropology* 24: 95-117.

—. 1998. *Ethnography through Thick and Thin*. Princeton, NJ: Princeton University Press.

Marcus, G., and M. Fischer. 1986. *Anthropology as Cultural Critique: An Experimental Moment in the Human Sciences*. Chicago: University of Chicago.

Markus, G., and S. Kitayama. 1991. "Culture and the Self: Implications for Cognition, Emotion and Motivation." *Psychological Review* 98: 224-53.

Marsh, G.P. 1864. *Man and Nature*. Trans. D. Lowenthal (1965). Cambridge, MA: Harvard University Press.

Marx, L. 1964. *The Machine in the Garden: Technology and the Pastoral America*. New York: Oxford University Press.

Maser, C. 1989. *Forest Primeval*. San Francisco: Sierra Club Books.

Mead, G.H. 1934. *Mind, Self and Society*. Vol. 1 (reissued 1967). Ed. C. Morris. Chicago: University of Chicago Press.

Melucci, A. 1994. "A Strange Kind of Newness: What's 'New' in New Social Movements." In E. Laraña, H. Johnston, and J. Gusfield, eds., *New Social Movements*, 101-30. Philadelphia: Temple University Press.

Merchant, C. 1992. *Radical Ecology*. New York: Routledge.

Michael, M. 1996. "Ignoring Science: Discourses of Ignorance in the Public Understanding of Science." In A. Irwin and B. Wynne, eds., *Misunderstanding Science?* 107-25. Cambridge: Cambridge University Press.

Milton, K. 1996. *Environmentalism and Cultural Theory: Exploring the Role of Anthropology in Environmental Discourse*. New York: Routledge.

—. 1999. "Nature Is Already Sacred." *Environmental Values* 8: 437-49.

—. 2000. "Animals, Personhood, and Emotion." Paper prepared for the 6th Biennial EASA Conference, Krakow, Poland, 26-29 July.

Murphy, K. 1999. "Judge Halts Pacific Northwest Timber Sales." *Vancouver Sun*, 4 August, D2.

Nader, L., ed. 1996. *Naked Science: Anthropological Inquiries into Boundaries, Power and Knowledge*. London: Routledge.

Naess, A. 1989. *Ecology, Community and Lifestyle*. Cambridge: Cambridge University Press.

Nash, R.F. 2001. *Wilderness and the American Mind*. 4th ed. New Haven, CT: Yale University Press.

Nelkin, D. 1987. *Selling Science*. New York: Freeman.

—. 1995. "Science Controversies: The Dynamics of Public Disputes in the United States." In S. Jasanoff, G. Murkle, I. Petersen, and T. Pinch, eds., *The Handbook of Science and Technology Studies*, 444-56. Thousand Oaks, CA: Sage Publications.

—, ed. 1992. *Controversies: Politics of Technical Decisions*. 3rd ed. Newbury Park, CA: Sage Publications.

Norse, E.A. 1990. *Ancient Forests of the Pacific Northwest*. Washington, DC: Island Press.

Norton, B. 1991. "Thoreau's Insect Analogies: Or, Why Environmentalists Hate Mainstream Economists." *Environmental Ethics* 13: 235-51.

Norwood, V. 1993. *Made from This Earth*. Chapel Hill: University of North Carolina Press.

Ortner, S.B. 1984. "Theory in Anthropology since the Sixties." *Comparative Studies in Society and History* 26(1): 126-66.

—, ed. 1999. Introduction to *The Fate of Culture: Geertz and Beyond*. 1–13. Berkeley: University of California Press.

Painter, J., Jr. 1992. "Man Sentenced in Log Theft Scheme." *Oregonian*, 30 November, C3.

Palmer, B. 1995. "National Survey on Forest Management." *Missouri Forest Management Notes* 7(1): 6-7.

Palsson, G., and A. Helgason. 1998. "Schooling and Skipperhood: The Development of Dexterity." *American Anthropologist* 100(4): 908-23.

Parish, S. 1994. *Moral Knowing in a Hindu Sacred City*. New York: Columbia University Press.

Peet, R., and M. Watts, eds. 1996. *Liberation Ecologies: Environment, Development, and Social Movements*. New York: Routledge.

Philibert, J.M. 1990. "The Politics of Tradition: Toward a Generic Culture in Vanuatu." In J.M. Philibert and F. Manning, eds., *Customs in Conflict: The Anthropology of a Changing World*, 251-73. Lewiston, NY: Broadview Press.

Pittman, A. 1993. "AFGA's New Director Faces Challenges." *What's Happening*, 10 June, 5.

Planalp, S. 1999. *Communicating Emotion: Social, Moral, and Cultural Processes*. New York: Cambridge University Press.

Porter, S. 1996. "The Wisdom of the Owl at Dusk: Cultural Conflict and Moral Disagreement over the Oregon Forests." PhD diss., Emory University, Georgia.

—. 1999. "The Pacific Northwest Forest Debate." *Worldviews: Environment, Culture, Religion* 3: 3-32.

Prigogine, I. 1997. *The End of Uncertainty*. New York: Free Press.

Proctor, J. 1996. "Whose Nature? The Contested Moral Terrain of Ancient Forests." In W. Cronon, ed., *Uncommon Ground: Rethinking the Human Place in Nature*, 269-97. New York: W.W. Norton and Company.

Pyle, R.M. 1996. *Where Bigfoot Walks: Crossing the Dark Divide.* Boston: Houghton-Mifflin.
Pyne, S. 1997. *Fire in America: A Cultural History of Wildland and Rural Fire.* Seattle, WA: University of Washington Press.
Quammen, D. 1998. "Planet of Weeds: Tallying the Losses of the Earth's Animals and Plants." *Harper's* 297(1781): 57-69.
Rappaport, R. 1968. *Pigs for the Ancestors: Ritual in the Ecology of a New Guinea People.* New Haven: Yale University Press.
Reddy, W. 1999. "Emotional Liberty: Politics and History in the Anthropology of Emotions." *Cultural Anthropology* 14(2): 256-88.
Redford, K. 1991. "The Ecologically Noble Savage." *Cultural Survival Quarterly* 15(1): 46-8.
Robbins, W. 1982. *Lumberjacks and Legislators.* College Station, TX: Texas A and M University Press.
—. 1985. "The Social Context of Forestry: The Pacific Northwest in the Twentieth Century." *Western Historical Quarterly* 16(4): 413-27.
—. 1988. *Hard Times in Paradise: Coos Bay, Oregon, 1850-1986.* Seattle: University of Washington Press.
—. 1989. "The Western Lumber Industry." In G.D. Nash and R.W. Etulain, eds., *The Twentieth-Century West: Historical Interpretations.* Albuquerque: University of New Mexico Press.
—. 1990. "The 'Luxuriant Landscape': The Great Douglas Fir Bioregion." *Oregon Humanities* Winter: 2-7.
—. 1997. *Landscapes of Promise.* Seattle: University of Washington Press.
Roberts, P. 1997. "The Federal Chain-Saw Massacre." *Harper's* 294: 37-51.
Robertson, L. 1994. "Labor Pains: The IWA Is Folding." *Register-Guard,* 24 February, B1-B2.
—. 1996. "Agents Study Station's Ruins: Fires." *Register-Guard,* 1 November, A1, A4.
Rodríguez, S. 1987. "Land, Water, and Ethnic Identity in Taos." In C. Briggs and J. Van Ness, eds., *Land, Water, and Culture.* Albuquerque: University of New Mexico Press.
Rolston, H. 1994. *Conserving Natural Value.* New York: Columbia University.
—. 1999. "Ethics and the Environment." In E. Baker and M. Richardson, eds., *Ethics Applied,* 407-37. New York: Simon and Schuster.
Rosaldo, M.Z. 1980. *Knowledge and Passion: Ilongot notions of Self and Social Life.* Cambridge: Cambridge University Press.
—. 1984. "Toward an Anthropology of Self and Feeling." In R.S. Shweder and R. Levine, eds., *Cultural Theory,* 137-57. Cambridge: Cambridge University Press.
Rosaldo, R. 1989. *Culture and Truth.* Boston, MA: Beacon Press.
—. 1994. "Race and Other Inequalities: The Borderlands in Arturo Islas's Migrant Souls." In Steven Gregory and Roger Sanjek, eds., *Race.* New Brunswick, NJ: Rutgers University Press.
Rouse, R. 1991. "Mexican Migration and the Social Space of Postmodernism." *Diaspora* 1(1): 8-23.
—. 1995. "Questions of Identity: Personhood and Collectivity in Transnational Migration to the United States." *Critique of Anthropology* 15(4): 351-80.
Satterfield, T.A. 2001. "In Search of Value Literacy: Suggestions for the Elicitation of Environmental Values." *Environmental Values* 10: 331-59.
Satterfield, T.A., and R. Gregory. 1998. "Reconciling Environmental Values and Pragmatic Choices." *Society and Natural Resources* 11(7): 629-47.
Satterfield, T.A., P. Slovic, and R. Gregory. 2000. "Narrative Valuation in a Policy Judgement Context." *Ecological Economics* 34(3): 315-31.
Scott, J. 1990. *Domination and the Arts of Resistance.* New Haven: Yale University Press.
Secretary of State. 2001. *Oregon Blue Book: 2001–2002.* Ed. T. Turgerson. Salem, OR: Office of the Secretary of State.
Shabecoff, P. 1993. *A Fierce Green Fire.* New York: Hill and Wang.
Shankman, P., and T.B. Ehlers. 2000. "The 'Exotic' and the 'Domestic': Regions and Representation in Cultural Anthropology." *Human Organization* 59(3): 289-99.
Sheridan, T. 1988. *Where the Dove Calls: The Political Ecology of a Peasant Corporate Community in Northwestern Mexico.* Tuscon: University of Arizona Press.
Shweder, R., and R. Levine, eds. 1984. *Cultural Theory: Essays on Mind, Self, and Emotion.* New York: Cambridge University Press.
Sillitoe, P. 1998a. "The Development of Indigenous Knowledge." *Current Anthropology* 39(2): 223-52.

—. 1998b. "What, Know Natives? Local Knowledge in Development." *Social Anthropology* 6(2): 203-20.

Slater, C. 1996. "Amazonia as Edenic Narrative." In W. Cronon, ed., *Uncommon Ground: Rethinking the Human Place in Nature,* 114-31. New York: Norton.

Snow, D.A., and R.D. Benford. 1992. "Master Frames and Cycles of Protest." In A.D. Morris and C. McClurg Mueller, eds., *Frontiers in Social Movement Theory,* 133-53. New Haven, CT: Yale University Press.

Snow, D., E. Burke Rochford, Jr., S. Worden, and D. Benford. 1986. "Frame Alignment Processes, Micromobilization, and Movement Participation." *American Sociological Review* 51: 464-81.

Sparks, A. 1990. "Little Folks Work." *Oregonian,* 1 September, D7.

Spirn, A. 1996. "Constructing Nature: The Legacy of Frederick Law Olmstead." In W. Cronon, ed., *Uncommon Ground: Rethinking the Human Place in Nature,* 91-113. New York: W.W. Norton and Company.

Stein, A. 2001. *The Stranger Next Door: The Story of a Small Community's Battle over Sex, Faith, and Civil Rights.* Boston, MA: Beacon Press.

Stiak, J. 1992. "Rubbernecking in Greentown." *What's Happening,* 2 July, 8-9, 32.

Stocker, M., and E. Hegeman. 1996. *Valuing Emotions.* New York: Cambridge University Press.

Strang, V. 1997. *Uncommon Ground: Cultural Landscapes and Environmental Values.* Oxford: Berg.

Strathern, M. 1980. "No Nature, No Culture." In C. MacCormack and M. Strathern, eds., *Nature, Culture, and Gender,* 174-222. New York: Cambridge University Press.

Strauss, C., and N. Quinn. 1997. *A Cognitive Theory of Cultural Meaning.* New York: Cambridge University Press.

Sullivan, H. 1956. *Clinical Studies in Psychiatry.* Ed. Helen Swick Perry, Mary Ladd Gawel, and Marth Gibbon. New York: Norton.

Szasz, A. 1994. *Ecopopulism: Toxic Waste and the Movement for Environmental Justice.* Minneapolis, MN: University of Minnesota Press.

Takacs, D. 1996. *The Idea of Biodiversity: Philosophies at Paradise.* Baltimore, MD: Johns-Hopkins University Press.

Taussig, M. 1980. *The Devil and Commodity Fetishism in South America.* Chapel Hill, NC: University of North Carolina Press.

Taylor, B. 1993. "Evoking the Ecological Self: Art as Resistance to War on Nature." *Peace Review* 5(2): 225-30.

—, ed. 1995. *Ecological Resistance Movements.* Albany: SUNY Press.

Taylor, C. 1992. *Multiculturalism and the Politics of Recognition.* Princeton, NJ: Princeton University.

Taylor, S. 1994. *Sleeping with the Industry.* Washington, DC: The Center for Public Integrity.

Terborgh, J. 1999. *Requiem for Nature.* Washington, DC: Island Press.

Thomas, K. 1983. *Man and the Natural World.* New York: Pantheon.

Tisdale, S. 1992. *Stepping Westward.* New York: Harper Perennial.

Tölölian. K. 1991. "The Nation State and Its Others: In Lieu of a Preface." *Diaspora* 1(1): 3-7.

Tyler, S. 1987. "The Vision Quest in the West." In Tyler, ed., *The Unspeakable: Discourse, Dialogue, and Rhetoric in the Post-Modern World,* 149-70. Madison: University of Wisconsin Press.

Vayda, A., and B. McKay. 1975. "New Directions in Ecology and Ecological Anthropology." *Annual Review of Anthropology* 5: 293-306.

Waller, D. 1996. "Friendly Fire: When Environmentalists Dehumanize American Indians." *American Indian Culture and Research Journal* 20(2): 107-26.

Walls, R.E. 1996. "Green Commonwealth: Forestry, Labor and Public Ritual in the Post-World War II Pacific Northwest." *Pacific Northwest Quarterly* 87(3): 117-29.

Waring, R., and J. Franklin. 1979. "Evergreen Coniferous Forests of the Pacific Northwest." *Science* 204: 1380-5.

Warren, K. 1988. "Toward an Ecofeminist Ethic." *Studies in the Humanities,* December, 140-56.

—. 1990. "The Power and Promise of Ecological Feminism." *Environmental Ethics* 12(2): 125-46.

Washington (State), Division of Forestry. 1912 and 1919. *Report of the State Forester.* Olympia, WA: Division of Forestry.

Watts, M., and R. Peet. 1996. "Conclusion: Towards a Theory of Liberation Ecology." In R. Peet and M. Watts, eds., *Liberation Ecologies: Environmental, Development, Social Movements*, 260-69. New York: Routledge.

Weeks, P. 1995. "Fisher Scientists: The Reconstruction of Scientific Discourse." *Human Organization* 54(4): 429-36.

White, G. 1990. "Moral Discourse and the Rhetoric of Emotions. In C. Lutz and L. Abu-Lughod, eds., *Language and the Politics of Emotion*, 46-68. New York: Cambridge.

—. 1992. "Emotions Inside Out: The Anthropology of Affect." Paper presented at the International Conference on Emotion and Culture, Eugene, OR, June.

White, R. 1980. *Land Use, Environment, and Social Change* (reissued 1992). Seattle: University of Washington Press.

—. 1991. *It's Your Misfortune and None of My Own: A New History of the American West.* Norman: University of Oklahoma Press.

—. 1996. "Are You an Environmentalist or Do You Work for a Living?" In W. Cronon, ed., *Uncommon Ground: Rethinking the Human Place in Nature*, 171-85. New York: W.W. Norton and Company.

Whitelaw, E. 1995. "Other Lessons from the Northwest." *Vermont Law Review* 19(2): 493-508.

Wierzbicka, A. 1994. "Cognitive Domains and the Structure of the Lexicon: The Case of Emotions." In L. Hirschfield and S. Gelman, eds., *Mapping the Mind*, 431-52. Cambridge: Cambridge University Press.

Wikan, U. 1990. *Managing Turbulent Hearts: A Balinese Formula for Living.* Chicago: University of Chicago Press.

Wolf, E. 1982. *Europe and the People without History.* Berkeley: University of California Press.

Wolf, R. 1993. "Water in the Evolution of National Forest Policy." Paper prepared for the Water Research Center. Tucson: University of Arizona.

Worster, D. 1993. *The Wealth of Nature.* New York: Oxford University Press.

Wynne, B. 1992. "Misunderstood Misunderstanding: Social Identities and the Public Uptake of Science." *Public Understanding of Science* 1: 281-304.

—. 1995. "Public Understanding of Science." In S. Jasanoff, G. Murkle, I. Petersen, and T. Pinch, eds., *The Handbook of Science and Technology Studies*, 361-88. Thousand Oaks, CA: Sage Publications.

Yearley, S. 1993. "Standing in for Nature: The Practicalities of Environmental Organizations' Use of Science." In K. Milton, ed., *Environmentalism: The View from Anthropology*, 59-72. New York: Routledge.

—. 1995. "The Environmental Challenge to Science Studies." In S. Jasanoff, G. Murkle, I. Petersen, and T. Pinch, eds., *Handbook of Science and Technology Studies*, 457-79. Thousand Oaks, CA: Sage Publications.

—. 1996. *Sociology, Environmentalism, Globalization.* London: Sage Publications.

Index

Notes: In page numbers, the italic letter "n" followed by a number indicates that the information is in a note. Many names for people, small towns and communities, activist groups, and some conferences are pseudonyms. They have been included in the index as given in the text for ease of reference.

Aboriginal people. *See also specific peoples*
  as activists, 56, 99, 103, 124, 177*n8*, 177*n9*
  authenticity, 105, 113, 124-26, 177*n9*, 177*n15*
  decimation of, 20, 172*n7*
  different cultures within, 102
  land use, current, 20, 103, 105, 127
  land use, traditional, 12-13, 19, 102-3, 105, 122-23, 130-33, 177*n8*
  relationship withWhite/Euro-American settlers, 99, 103-4, 107-8, 115, 129-30, 176*n1*
  removal to reservations, 26, 130, 173*n17*
  rights, 127, 178*n19*
  romanticized, 19, 99-103, 105-6, 108, 113, 115, 118, 127-28, 176*n2*, 177*n4*, 177*n8*, 177*n11*
  undermined by activists, 120-26, 131-32, 134
  used by activists, 107-8, 115, 117-21, 126-27, 133-34
aesthetics, 18
AFGA. *See* Ancient Forest Grassroots Alliance
AFL-CIO, 32
Agriculture. *See* farming
Alan (environmentalist), 51-52
Alaska, 39
alder, 23, 47, 175*n13*
Amazon rain forest, 125-26
American Federation of Labor and Congress of Industrial Organizations (AFL-CIO), 32
Ancient Forest Grassroots Alliance (AFGA)
  directors, 65, 106

  growth, 41, 174*n7*
  and research for this book, 41
  on science, 85
  subgroups, 150
  women members, 174*n8*, 175*n19*
Ancient Forest Movement. *See also* environmentalists
  annual conferences, 59
  emergence, 11, 40-41
  growth, 3
anger, 141-43, 147, 158, 166
animal species. *See also* endangered species; *specific animals*
  game, 19
  indicator species, 16-17
  in snags and deadfall, 16
anthropology
  on Aboriginal "nobility," 100, 176*n2*, 177*n4*
  favours "exotic" over "domestic," 102, 177*n6*
  and primitivism, 101-4
  study of emotion, 136-37
  study of environmental conflict, 162-63, 180*n4*
anti-poverty movement, 177-78*n17*
anti-racism movement, 177-78*n17*
Armstrong, Carolyn, 68
Audubon Society, 41, 43, 71, 174*n9*
Austin, Mary, 26

Baker, Arlene, 125-26, 138-39, 141
Baker, Chris, 138-39, 142-43, 152
Baker, Gary, 138
Balinese people, 178*n2*
Basso, K.H., 10

Bend, Oregon, 41
Benedict, Ruth, 101-2
Berkhofer, R.F., 122
Bill (logger), 45
biodiversity, 18, 25-26, 33-34
biomass, 16, 18, 172*n3*
Bohm, David, 84
Boise-Cascade (company), 31, 173*n20*
Border, Jim, 116
Bourdieu, P., 176*n13*, 177*n13*
Bourgois, P., 178*n22*, 180*n8*
Brian (environmentalist), 53-54
British Columbia, 3, 10, 158
Brookside, Oregon, 91
Brosius, P., 170
Buege, D.J., 105, 126, 177*n9*
Bunyan, Paul, 32, 74, 173*n22*
Burns, David, 87-88, 107
Bush, George H.W., 35, 42
Bush, George W., 36, 39

Calapooia people, 19, 124
Calapooia Valley, Oregon, 172*n6*
Caledon, Oregon, 71-72, 74, 91, 125,
    175*n4*
California, 3, 10, 31, 33
Carr, Jeanne, 173*n16*
Carson, Rachel, 33, 37
Carver, Nathan, 108
Cascade Mountains, 5, 9-10, 131-32
Caterpillar (company), 4, 172*n2*
Caufield, Catherine, 17
caulk boots, 175*n17*
Chinook people, 124
chokers, 44
Clark and Lewis, 122, 130, 177*n11*
class
    and emotionalism, 137, 142, 147
    environmentalists, 168, 177-78*n17*,
        180*n3*
    importance to cultural conflict, 168
    loggers, 168, 178*n24*, 180*n3*
clear-cutting
    environmental impacts, 16, 23-25,
        173*n13*
    and science, 22, 88-89, 91-92
Cleveland, Grover, 65
Clinton, Bill
    elected, 32
    endorses need for science, 12, 80-81, 93
    environmental lobbying, 39, 56
    Forest Plan, 35-36, 64-65, 75, 81, 129,
        174*n28*, 175*n1*
    Forest Summit, 35, 41, 80, 138, 165,
        175*n1*
Coast Salish people, 124
coho salmon, 56
common sense
    definitions, 90-91

and science, 90, 92
communities
    and culture, 108-10, 124
    and depletion of resources, 66
    of environmentalists, 112-15
    others' identification, 69
    and place, 10, 108, 124
    self-identification, 38, 75, 106
conflict as central to ethnographic study,
    162, 165, 180*n2*
conifers, 15-17, 21
Conklin, B.A., 115
conservation ethic, 11, 27-28
Coos Bay, Oregon, 22
Corvallis, Oregon, 40, 172*n2*
cost-benefit analyses, 169-70
Costas, Michael, 85, 117-18, 145
cultural pluralism, 101, 178*n26*
cultural relativism, 102, 177*n5*
culture
    authenticity, 108, 113, 115, 133-34,
        177*n15*
    definitions, 3-4, 6, 165
    degrees of, 178*n26*
    and ecological nobility, 102-6, 133-34
    and emotion, 147, 178*n2*, 179*n11*
    fire as metaphor, 132
    generating discourse, 164, 166-67
    and identity, 6-7, 12, 75-76
    and loggers, 39, 63-64, 66-68, 75, 109-
        10, 127-29
    and place, 105, 108, 110-12, 120, 177*n9*
    and power, 6-8, 12, 100, 115, 165
    resistant forces vs. productive forces,
        167-69
    and science, 81
    zones of conflict, 165-66

Dan (logger), 45
Darwin, Charles, 26
Dawson, Julie, 86-87, 107, 109-10
DDT (dichlorodiphenyltrichloroethane),
    33
deadfall, 16
DeFazio, Peter, 4, 36
dialogues
    between subordinate groups, 162
    of identity, 8-10, 66-67
    oppositional, 9
dichlorodiphenyltrichloroethane (DDT),
    33
Dietrich, William, 17
discourses
    dominant, 7, 161, 164
    ecological nobility, 120-21, 126-27,
        133-34
    interplay of, 7-8, 161, 163-64, 167
    and policy making, 169-71
    and power, 180*n6*

scientific, 81-82, 97-98
and stigma, 64, 76-77
subordinate, 7, 164
used to emphasize human links, 69
Donation Land Laws (US, 1850), 20
Doug (logger), 46-51
Douglas fir, 15-17, 34
Durbin, Kathie, 59
Dwyer, William, 34-36, 39

ecofeminism, 58-59, 144-45, 179n8
ecological nobility
  environmentalists' views, 102-6, 108, 133
  loggers' views, 120-22, 124-26, 133
ecological resistance movements, 12
ecosystem management
  history, 25-26, 34-35
  through human mortality, 117
Einstein, Albert, 90
Elger, Edward, 90
Elgin, Frank, 18
elk, 47
embarrassment, 179n14
Emergency Supplemental and Rescissions
  Act (US, 1995), 36
Emerson, Ralph Waldo, 25, 173n16
emotion
  anger, 141-43, 147, 158, 166
  and culture, 147, 178n2, 179n11
  definitions, 140
  embarrassment, 179n14
  and environmentalists, 13, 86-89, 138-
    39, 142-47, 153-56, 166, 179n17
  fear, 13-14, 150-54, 156-57, 159, 166
  and identity, 140-41, 147-48
  as lens to study behaviour, 136-37
  and loggers, 137-39, 141-43, 147,
    156-59, 166
  and men, 145-47
  and morality, 148-49, 153-54, 158-59
  paired with science, 86-89, 146-47
  and power, 137, 149
  used by activists, 13, 86-89, 135-36, 141,
    148-49, 153-59
  value of controlling, 139-41
  and women, 137, 139-41, 143-45, 147
endangered species
  Clinton administration, 35-36
  grizzly bear, 158
  legislation, 5, 34-36, 175n1
  normal systemic function, 86
  northern spotted owl, 3-4, 6, 16-17, 32-
    35, 42-43, 45-46, 52, 65, 69, 82,
    92-93, 110, 158
  role of FWS, 173n25
  salmon, 20, 56, 158
Endangered Species Act (US, 1973), 5, 34-
  36, 175n1

environmental impact statements, 34
environmentalists. *See also* Ancient Forest
    Movement
  and class, 95, 168, 174n8, 177-78n17,
    180n3
  cultural legitimacy, 12-13, 108-10, 133
  definition of forest, 83
  disparate views of, 4-5, 28-29
  emotionalism, 13, 86-89, 138-39, 142-
    47, 153-56, 166, 179n17
  as grassroots activists, 11, 63-64
  "green imperialism," 178n19
  history, 24-29, 33, 173n14
  loggers' view of, 65, 96, 121-22, 129,
    152-53
  men as, 58
  nostalgia, 106-8
  perception of forest value, 15, 18, 59
  and place, 112-15, 134, 177n14
  in policy making, 165
  protest activities, 3, 51-55, 78-80
  relationship with nature, 18, 112-15,
    154-56, 160-61
  and research for this book, 41, 51-5
  and science, 8, 12, 81-90, 94, 97, 176n5
  use of fear, 13-14, 150-51, 153-54
  view of Aboriginal people, 104, 106-8,
    115-16, 118, 127
  view of loggers, 64-70, 77, 108-12
  view of logging injunctions, 5, 7
  women as, 58-59, 144, 174n8, 175n19
environmental justice movement, 177-
  78n17
Eric (Oregon student), 4
Eugene, Oregon, 41, 73
Evans, Glen, 151-52
Evans, Mark, 57-58, 76, 122-23, 151-52,
  178n21

farming, 20-21, 172n9
fear
  used by environmentalists, 13-14, 150-
    51, 153-54
  used by loggers, 13-14, 151-53, 156-57,
    159, 166
Federal Lands and Forest Health Protection Act (US, 1995), 35-36
Fenimore Cooper, Susan, 25-26
fire
  1994 wildfires, 35
  effect on species, 16
  in land management, 19, 118, 122-23,
    131-32
  logging-related risks, 24, 45-46, 173n13
floods, 23, 25
forest
  definitions, 8, 83, 89, 92, 132
  fragmented, 24

multilayered canopy of, 15
potential depletion of, 18, 24
for public use, 27-28
Forest Community Movement. *See also*
    loggers
    emergence, 11, 38
    ethos, 39
    growth, 3
Forest Congress (1895), 28
foresters, 29
forest management
    based on natural processes, 123, 131-32
    history, 24, 27-28, 33, 35-37
    policy making, 162, 165, 169-71
    and science, 81, 83, 85, 88-89, 91-92
Forest Management Act (US, 1897), 28
forest products, 29-31
Forest Service. *See* United States Forest
    Service
Franklin, Jerry, 18, 172*n2*
Fuller, Steve, 95-97, 122, 127
fungus, 17, 82
FWS. *See* United States Fish and Wildlife
    Service

gardening, 25-26
Gautier, Ronald, 19, 131-32, 138
Geertz, Clifford, 91-92, 101
gender
    and emotionalism, 137, 139-41, 143-47
    and environmentalism, 58-59, 144,
        174*n8*, 175*n19*
    and human-nature relationship, 143-45,
        179*n7*
    and timber activism, 174*n8*
General Land Revision Act (US, 1891,
    1897), 26
Georgia-Pacific (company), 31
Gezon, L., 162
Gifford Pinchot National Forest, 61
Goffman, Erving, 72, 75-76
gold rush, 172*n8*
Gore, Al, 36
Grant, Martha, 94
grassroots movements
    competition for activist credentials,
        11-12, 63-67, 69-70, 77
    discontent with national groups, 41
    emergence, 11
    environmentalist conferences, 55-59
    identity seeking, 38, 75-76
    supported by GLA, 32
GreenWorld, 34
grizzly bear, 158
Gyppo Loggers Association (GLA), 32, 39

Hanel Lumber, Oregon, 40
Hardesty, Jean, 129

Hardesty, Stu, 129-31
Harrison, Benjamin, 26
harvest
    restrictions, 35-36, 64, 128-29
    and species diversity, 83
Hatfield, Mark, 78
Hawkins, Bill, 74-75, 132
Headland, T., 104
hemlock, 16, 23
Hetch Hetchy Valley, Yosemite National
    Park, 28-29
Hill, Gordon, 150-51
Hochschild, A.R., 137
Holland, B., 142, 176*n13*
Homestead Act (US, 1863), 21
housing, 15-17
human-nature relationship
    Aboriginal people, 20, 100, 105-6, 116,
        120, 122-23, 177*n8*
    effect on society, 163
    emotional-spiritual, 143-47, 149, 153-
        54, 165, 172*n1*, 179*n15*, 180*n6*
    environmentalists, 18, 112-15, 154-56,
        160-61
    history, 20, 24-25, 29, 87, 172*n3*
    and humility, 57-58, 87-88
    ideal, 116-17
    loggers, 15, 17-18, 24, 47, 49, 59, 144,
        156, 160, 179*n6*
    men, 144
    primitive cultures, 104
    women, 143-45, 179*n7*
hunting, 52-53

Ian (timber activist), 124
identity
    construction, 9, 38, 96
    and culture, 6-7, 12, 75-76
    definitions, 8
    dialogues, 8-10, 66-67, 72
    and emotion, 140-41, 147-48
    and experience, 96
    and grassroots movements, 38, 63-64,
        69-70, 75-76
    and place, 9-10
    and resistance to change, 167
    rights as expression of, 63, 72, 76,
        162
Ifaluk people, 148-49
International Woodworkers of America
    (IWA), 32
International Workers of the World
    (Wobblies), 23, 32
Inuit, 178*n2*
Italians, 178*n2*
ITT-Rayonier (company), 31
IWA (International Woodworkers of
    America), 32

James (environmentalist), 53-54
James, William, 90, 140, 178*n3*
Janz, Charlie, 40
Jarmon, Marie, 65-67, 104
Jasper, J., 148

Kaley, Rita, 40
Karuk people, 56
Katrina (Oregon student), 5
Kayapo people, 115
knowledge
 and common sense, 90-92
 experiential vs. theoretical, 94-98
Kohn, Kathryn, 18
Kramer, Frank, 128
Kristin (Oregon student), 5

Ladner, Monica, 107, 143-44
land ethic, 11, 29-30
land use
 Aboriginal people, 12-13, 19-20, 102-3,
  105, 122-23, 127, 130-33, 177*n8*
 recreational, 30, 33, 129, 143
 White/Euro-American settlers, 19-23, 67,
  122, 172*n9*, 176*n8*
Lange, J., 167-68
Langston, Nancy, 177*n12*
legislation
 Donation Land Laws (US, 1850), 20
 effect on timber extraction, 31
 Emergency Supplemental and
  Rescissions Act (US, 1995), 36
 Endangered Species Act (US, 1973), 5,
  34-36, 175*n1*
 Federal Lands and Forest Health
  Protection Act (US, 1995), 35-36
 Forest Management Act (US, 1987), 28
 General Land Revision Act (US, 1891,
  1897), 26
 Homestead Act (US, 1863), 21
 Multiple Use Sustained Yield Act (US,
  1960), 33, 128, 173*n23*
 National Environmental Policy Act (US,
  1970), 34, 175*n1*
 National Forest Management Act (US,
  1976), 5, 34, 175*n1*
 National Park Act (US, 1890), 28
 Preemption Act (US, 1841), 20-21
 Timber and Stone Act (US, 1878), 21
 Wilderness Act (US, 1964), 33, 128
Leopold, Aldo, 11, 29-30, 33, 37, 55, 86,
 119
Lewis and Clark, 122, 130, 177*n11*
Lincoln, B., 69
log-scaling trials, 172-73*n10*
Loggers. *See also* Forest Community
 Movement; timber activists

and class, 168, 178*n24*, 180*n3*
clothing and equipment, 44, 46, 53,
 175*n17*
cultural legitimacy, 12-13, 63-64, 75, 77,
 106, 109-12, 127-32, 134
disparate views of, 4-5, 73
emotional control, 137-39, 141-43, 147,
 156-59, 166
environmentalists' view of, 64-70, 77,
 108-12
as grassroots activists, 63-64, 75
gyppo, 31-32, 121
job losses, 4, 35, 71, 111-12, 124, 172*n1*
labour movements and organizations,
 23, 32, 175*n18*
and place, 110-12, 177*n14*
in policy making, 165
protest activities, 6, 68
relationship with nature, 15, 17-18, 24,
 47, 49, 59, 144, 156, 160, 179*n6*
and research for this book, 42-51, 57-58,
 71-72, 74, 95
rights, 63-64, 72, 128-29
romanticized, 32, 67-68, 73
and science, 8, 12, 86, 90-97
stigmatized, 43, 70-77, 137, 142
use of fear, 13-14, 151-53, 156-57, 159,
 166
view of Aboriginal people, 106, 120-28
view of environmentalists, 65, 96, 121-
 22, 129, 152-53
view of logging injunctions, 5, 7
women as, 174*n8*
worklife, 43-51
logging. *See also* timber industry
 clear-cutting, 16, 22-25, 61-62, 88-89,
  91-92, 173*n13*
 defined, 44
 equipment and technology, 44, 50-52, 61
 harvest restrictions, 35-36, 64, 128-29
 high-lead, 23
 legitimized by science, 8, 86, 92
 roads, 42, 47, 92, 112
 salvage operations, 35-36, 175*n5*
 selective cutting, 22, 173*n11*
 technological history, 21-24, 30,
  173*n11*, 173*n12*
 truck blockades, 3, 6, 54
logging injunctions, 5, 7, 34-36, 39-40
logging roads, 42, 47, 92, 112
Louisiana Pacific (company), 31
Lujan, Manual, 42
Lutz, Catherine, 148-49, 178-79*n4*
Lyons, Warren, 113

Madagascar, 162
Malaysia, 170

mammals
  as game, 19
  grazing on infant forest, 24, 47, 123
  small, 17, 158
Marcus, G., 162
Marden, Dan, 156-57
Marsh, George Perkins, 24-25, 27, 37,
  173*n15*
Marx, Leo, 176*n8*
Masai people, 120
Mashpee people, 105
Mason, Beverly, 74-76, 91-92, 129,
  176*n12*
Mason, Vern, 91-92
Mason Logging, Oregon, 71, 74, 92
Massachusetts, 34
McKinley, William, 27
media, 65, 124-25, 135-36
men
  emotionalism, 145-47
  as environmentalists, 58
  relationship with nature, 144
Michael (environmentalist), 51
Michael (Oregon student), 5
Miley, Debbie, 40
Millennium Grove, Oregon, 40, 174*n5*
mills, 21, 31, 66
Milton, Jeff, 118, 122
morality
  of concern for humans, 179*n6*
  of concern for nature, 179*n6*
  and emotion, 148-49, 153-54, 158-59
  and fear, 13-14, 150-53
  of nature, 25
Mott, Cheryl, 124-25
Mott, Stacey, 124
Mt. Hood National Forest, 78
Mountain, Lila, 56-57
mudslides, 23
Muir, John, 11, 25-29, 37, 173*n16*
Multiple Use Sustained Yield Act (US,
  1960), 33, 128, 173*n23*
mycorrhizal fungus, 17, 82

National Environmental Policy Act (US,
  1970), 34, 175*n1*
National Forest Management Act (US,
  1976), 5, 34, 175*n1*
national forests
  acreage, 26-27
  history, 25-28
  in Oregon, 30
  U.S. national guidelines, 28
National Park Act (US, 1890), 28
national parks, 25-29, 173*n17*
natural resource workers, 12, 165. *See also*
  specific types

nature
  definitions, 163-65
  moral goodness, 25
  relationship with humans (*see* human-
    nature relationship)
  robustness, 47
  scientific views, 84-88
Neah Bay, Washington, 127
New Mexico, 162
Newman, Doug, 95
Nixon, Richard, 34
non-governmental organizations, 41
Norman, Greg, 85, 107, 146-47
northern spotted owl
  battleground, 3-4, 69
  biological corridor, 52
  Clinton plan, 65
  emergence as concern, 33-35, 82
  vs. human needs, 6, 158
  as indicator species, 16-17
  lack of effect of labour unions, 32
  loggers' view of, 42-43, 45-46, 110
  reserves, 34
  Roseburg hearing, 42-43
  scientific evaluations, 92-93
Northwest Forest Plan (US, 1994), 35-36,
  64, 75, 81, 174*n28*

OFCC. *See* Oregon Forest Community
  Coalition
oil, 39
old-growth forests
  biological diversity, 16
  decay cycle, 17
  definitions, 15-16, 45
  diverse perceptions, 15, 17-18, 45, 59
  ecological function, 172*n2*
  Oregon battlefield, 3-6
  pictures, 60
  remaining, 30
Oregon
  1990s political climate, 41-42
  annual cut, 30, 36
  celebration of loggers, 32, 73
  culture of rural life, 108-9
  early exploration, 122, 130
  environmentalists' protest activities,
    51-55
  gay and lesbian population, 57, 76
  gold rush, 172*n8*
  job losses in timber industry, 35
  old-growth forests, 3-6
  public land, 30-31
  settlement, 19-21, 122, 172*n8*, 172*n9*
  Squaw Three Timber Sale, 40
  timber industry history, 20-31
  wilderness reserves, 33

Oregon and California Railroad, 31
Oregon Eco-Defenders, 51
Oregon Forest Community Coalition
    (OFCC)
    defends grassroots status, 40
    at environmentalists' conference, 57
    membership, 39-40, 71, 94, 174*n3*,
        174*n4*
    picture, 79
    and research for this book, 41, 124
Ortner, S.B., 165, 180*n7*
Owen, Lorna, 71-72, 76, 92-93
Owl. *See* northern spotted owl

Paine, Thomas, 90
partnership model, 179*n7*
Paul (Oregon student), 5
Peet, R., 164
Pinchot, Gifford, 11, 27-29, 37, 89
place
    and communities, 10, 108, 124
    and culture, 105, 108, 110-12, 120,
        177*n9*
    and environmentalists, 112-15, 134,
        177*n14*
    and identity, 9-10
    importance, 177*n14*
policy making
    and activists' dialogue, 165
    decision making and conflict, 162, 169
    power of language, 169-70
    stakeholder value elicitation, 169, 171
Portland, Oregon, 16, 31, 41, 80, 175*n1*
power
    of big industry, 31
    and culture, 6-8, 12, 100, 115, 165
    and discourse, 180*n6*
    and emotion, 137, 149
    grassroots, 64, 66
    and policy making, 169-70
    and stigmatization, 64, 72, 76-77
    symbolic, 12, 107
    of unions, 32
Preemption Act (US, 1841), 20-21
privatization, 38-39
protest activities
    by environmentalists, 3, 51-55, 78-80
    by loggers, 6, 68
    Oregon Eco-Defenders, 51-55
public lands
    biodiversity management, 33-34
    preservation history, 25-28
    privatization for resource extraction, 38-
        39
    public participation, 34, 169, 171
Puget Mill Company, 21
Puget Sound, Washington, 21, 172*n7*
Pyle, Robert Michael, 88

Quammen, David, 116

racism, 121, 126, 177-78*n17*, 178*n23*
railroads, 22-23, 31
rain forests
    Amazon, 125-26
    Oregon, 3-6
ranger stations, 3
Rappaport, R., 101
Reagan, Ronald, 32, 38
Reddy, W., 139
Redford, K., 103, 177*n7*, 177*n8*
Richardson, Donald, 146, 153-56, 179*n10*,
    179*n17*
Rick (logger), 44-46
rights
    as expression of identity, 63, 72, 76, 162
    of loggers, 63-64, 72, 128-29
    of women, 144
Robbins, William, 22
Rodríguez, S., 162
Rolston, H., 155
romanticism
    of Aboriginal people, 19, 99-103, 105-6,
        108, 113, 115, 118, 127-28, 177*n2*,
        177*n4*, 177*n8*, 177*n11*
    of loggers, 32, 67-68, 73
    of science, 84
Roosevelt, Theodore, 11, 27-29, 37, 89
Rosaldo, Michelle, 179*n11*
Roseburg, Oregon, 42
Rousseau, Jean-Jacques, 101, 177*n4*

Sagebrush Rebellion, 38
St. Augustine, 55
salmon, 20, 56, 158
San Francisco, 28-29
Sanderson Brothers, Oregon, 46
Save Our Community (SOC), Oregon,
    71-72, 74-75, 91-92, 125
science
    and common sense, 90-92
    and culture, 81
    definition of forest, 83
    different views of, 12, 81-83, 94-98
    environmentalists' use of, 8, 12, 81-90,
        94, 97, 176*n5*
    expert vs. non-expert views, 81-82, 92-
        93, 176*n14*
    interpretations of nature, 84-88
    loggers' use of, 8, 12, 86, 90-97
    mistrust of, 88-90, 92-94
    and mystery, 84-85, 97
    and northern spotted owl, 92-93
    in Northwest Plan, 12, 80-81, 93
    paired with emotion, 86-89, 146-47
    and praxis, 94-97
    romanticism of, 84

second-growth forests, 62, 92
self
    authoring, 147-48, 179*n5*
    definitions, 154-56, 179*n13*
    and emotion, 147-48, 153-54, 179*n14*
Shabecoff, Philip, 24
Sierra Club, 25, 41, 173*n23*
Silver Fire Round-Up, Oregon, 175*n5*
Silver Valley Timber Action Coalition, 94
Simon, Andrew
    ambivalence towards science, 84, 89-90
    attempt to establish place viability, 112-15
    imagines the future, 118-20
    on loggers as victims, 67-68
    on nobility of Aboriginal people, 108
Sinai-Bedouin people, 117
Sioux people, 124
Smithers, Andrea, 57
snags, 16, 152
social movements
    competition for public support, 12
    creation, 9, 38, 70
    synthesis into environmental justice
        movement, 177-78*n17*
soil erosion, 23, 25
Sonny (logger), 48, 50
Spirit of the Earth Forest Activist Conference (1993), 55-59
spirituality
    of forests, 18
    in human relationship with nature, 143-47, 149, 153-54, 165, 172*n1*, 179*n15*, 180*n6*
    of wilderness, 25-26
spotted owl. *See* northern spotted owl
Squaw Three Timber Sale, 40
stigma
    defined, 70
    and loggers, 43, 70-77, 137, 142
    natural history, 73-75
    promotes pride, 75-76
    and social isolation, 72
    symbolic language, 74-75, 153
    as tool for political recognition, 64, 76-77
Stratton, Jim
    assists author in research, 43, 46, 175*n11*, 175*n12*
    on dangers in logging, 151-52
    on environmental impact of logging, 23
    on stigmatization of loggers, 43, 70-71
    uses fire as metaphor, 132
Sumner Mountain, Oregon, 112-15, 120
surveys, 169-70

Takielma people, 56
Taylor, Charles, 35, 76, 175*n6*
Tess (environmentalist), 51-52, 54-55

Thomas, Jack Ward, 175-76*n3*
Thoreau, Henry David, 25, 55
timber activists. *See also* loggers
    avoid emotional language, 13
    cultural legitimacy, 12-13, 77
    definition of forest, 83
    as grassroots activists, 11, 64
    and research for this book, 41, 174*n8*
    and science, 12, 81-83, 86, 90-93, 176*n5*
    stigmatization of, 71
    use of science, 90-93
    views of science, 94-97
Timber and Stone Act (US, 1878), 21
timber industry. *See also* logging
    "big" vs. "small," 31, 40
    history in Pacific Northwest, 20-31
    perception of forest value, 17-18, 27-28
    supported forest reserves, 26, 28
    workforce, 4, 23, 30-32, 35, 65-67, 77, 172*n1*
Tongass National Forest, 89
Treat, Mary, 26
tree sitting, 54, 174*n6*
tree spiking, 3
trees
    age and biomass, 15, 172*n3*
    climax species, 16
    coniferous, 15
    dead, 16, 152
truck blockades, 3, 6, 54
Tualatin Valley, Oregon, 19
Tuolumne River, Hetch Hetchy Valley, 29
Tyler, Stephen, 176*n9*

unions
    division with environmentalists, 56, 175*n18*
    international movements, 23
    memberships, 32
    reduced power, 32
    supported Clinton, 32
United States
    Department of Agriculture, 27-28, 30, 36, 173*n25*
    Department of the Interior, 27, 30, 173*n25*
    National Academy of Science, 28
United States Bureau of Land Management, 5, 30-31, 169, 173*n19*, 173*n25*
United States Environmental Protection Agency, 169
United States Fish and Wildlife Service, 34, 169, 173*n25*
United States Forest Experiment Station, Corvallis, 172*n2*
United States Forest Service
    biodiversity management, 31, 33-34, 92, 169, 173*n19*, 177*n12*

established, 27
internal conflict, 86, 173*n25*
logging roads, 42, 47, 92, 112
Millennium Grove dispute, 174*n5*
nickname for employees, 52, 175*n15*
Old-Growth Definition Task Group, 15
policy making, 169
reaction to environmental protests, 52
Sierra Club's suspicion of, 173*n23*
timber sales, 5, 30, 35-36, 125
under Agriculture, 27-28, 36, 173*n25*
whistle blowers, 86
wilderness reserves, 33
United States Forest Summit (1993), 35,
   41, 80, 138, 175*n1*
Utku people, 178*n2*

Vermont, 173*n15*
Vietnam, 150-51

Waite, Matthew
   about, 64-65
   on emotionality of old-growth contro-
      versy, 150
   on loggers' unrootedness, 110-12
   makes sexist "joke," 58
   on timber worker as addict, 68-69
   on timber worker as pawn, 65-66, 69
   soundbites, 55, 65, 175*n3*
Walls, Peter, 67
Walton, Bill, 121-22
Washington State, 3, 23, 30, 33
Watt, James, 38
Watts, M., 164
WCIW (Western Council of Industrial
   Workers), 32
West, Elliot, 131
Western Council of Industrial Workers
   (WCIW), 32
Weyerhaeuser (company), 18, 31, 173*n20*
Weyerhaeuser, Frederick, 23
whales, 127

White, Richard, 94, 176*n13*, 177*n1*,
   178*n11*, 178*n20*
White/Euro-American people
   and emotionalism, 137
   Oregon settlement, 19-21, 122, 130-31,
      172*n8*, 172*n9*
   population boom in Pacific Northwest,
      23
   relationship with Aboriginal people, 99,
      131, 176*n1*
   settlers and land use, 19-23, 67, 122,
      172*n9*, 176*n8*
wilderness
   loggers' views, 128, 132
   preservation, 25-26, 28-30, 33
Wilderness Act (US, 1964), 33, 128
Wilderness Society (US), 29-30, 41
wildlife reserves, Alaska, 39
Willamette Industries, 31, 174*n5*
Willamette National Forest, 46-47, 62, 95
Willamette Valley, Oregon, 19, 122-23,
   172*n6*
Williams, Terry Tempest, 55
Willings, Laurie, 72-73
Wilson, Paul, 106, 175-76*n3*
Wise Use Movement, 38-39
Wobblies (International Workers of the
   World), 23, 32
women
   emotionalism, 137, 139-41, 143-45, 147
   as environmentalists, 58-59, 144, 174*n8*,
      175*n19*
   as loggers, 174*n8*
   relationship with nature, 143-45, 179*n7*
   rights, 144
wood products industry, 5
Wynne, Brian, 96

X-L Timber, Oregon, 40

Yosemite National Park, 28-29, 173*n17*